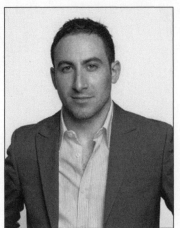

Mark Mahaney

About the Editor

ADAM BLY is the founder of *Seed*. He has lectured at the World Economic Forum, the Museum of Modern Art, the Royal Society, the National Academy of Sciences, and the Academy of Sciences for the Developing World, as well as at Harvard and Peking University on the future of science and its changing role in society. He is the recipient of the Golden Jubilee Medal from Queen Elizabeth II and was named a Young Global Leader by the World Economic Forum. He grew up in Montreal, Canada, where he worked in a lab studying the biochemistry of cancer, and now lives in New York.

Science
Is
Culture

Science Is Culture

Conversations at the New Intersection of Science + Society

Edited and with an Introduction by Adam Bly

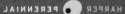

Science Is Culture

Conversations at the New Intersection of Science + Society

Edited and with an Introduction by Adam Bly

HARPER 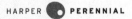 PERENNIAL

NEW YORK • LONDON • TORONTO • SYDNEY • NEW DELHI • AUCKLAND

HARPER ⬤ PERENNIAL

Photographs appearing in chapter 1 and 2 by Henry Leutwyler. All other photographs by Julian Dufort.

HarperCollins books may be purchased for educational, business, or sales promotional use. For information please write: Special Markets Department, HarperCollins Publishers, 10 East 53rd Street, New York, NY 10022.

FIRST EDITION

Designed by Justin Dodd

Library of Congress Cataloging-in-Publication data is available upon request.

ISBN 978-0-06-183654-1

10 11 12 13 14   OV/RRD   10 9 8 7 6 5 4 3 2 1

Contents

Introduction

When I first jotted "science is culture" down on a mock-up of the cover of *Seed*, it was more of an aspiration than an observation. I started a science magazine during an American administration that not only opposed science but devalued empiricism. This emboldened activists around the world who rejected evolution and sought to reform science education with their own version of truth. But at the same time, in the vitality of this antiscientific movement, you could sense the imminent rise of a cultural shift they sought to prevent. I had the idea that science was becoming the fundamental driver of our times and held unique promise to improve the state of the world. I believed that we were on the cusp of a twenty-first-century scientific renaissance, and I wanted to create a magazine to capture this cultural shift. I called it *Seed* because that's how I saw science: where things start, beneath the surface of all that surrounds us.

Today our tagline seems self-evident: science *is* culture. In the last decade, science has transformed the social, political, economic, aesthetic, and intellectual landscape. It is reshaping our understanding of who we are and where we come from and modernizing our system of values—how we regard our planet and one another. Other forces undeniably affect the state of the world—faith, democracy, and free markets among them. But science is the overwhelming and universal agent of change. Today, science affects every single person on the planet.

These are times for great optimism—about what we can know and accomplish. In the last decade, we mapped the human genome and the cosmic microwave background. We confirmed the existence of dark energy and the age of the universe, and know that it is expanding. We sequenced the rice genome. We found evidence of oceans on Mars and landed for the first time on another planet's (Saturn) moon. We know that we have more in common with Earth's other biota than we once thought. We built a machine capable of approximating the big bang, used the Internet to study human behavior, and produced more data than in all prior human history. We awarded scientists the Nobel Peace Prize.

These are also times of great uncertainty. We now notice that the systems that support life and commerce on this planet are riddled with complexity and interdependence of an unimaginable scale. We see that we cannot solve epidemics, for example, with solving climate change; we cannot solve climate change without rethinking economic growth; we cannot reconsider growth without an understanding of population and demographics; and we cannot anticipate population shifts without considering epidemics. The toolbox that served us well in the twentieth century is inadequate to wrestle with the systemic challenges apparent at the dawn of the twenty-first.

We need a new way of looking at the world. We need to rekindle the conviction that knowledge is good, and more knowledge is better. We need to be reminded that disciplines are man-made creations, useful only to a degree. Like the Renaissance of the fifteenth and sixteenth centuries, the renaissance before us will be characterized by a revolution in how knowledge is gathered, synthesized, and applied to society. And once again, science will mean more to us than its output, than drugs and technologies; more than something to be governed, science will become a way of governing and thinking.

Today, we think scientifically about issues like health and the environment, but soon we will also think scientifically about the size of cities, the reason for poverty, the basis of morality, the resilience of markets, and much more.

Science is a methodology and philosophy rooted in evidence, kept in check by persistent inquiry, and bounded by the constraints of a self-critical and rigorous method. Science is a lens through which we can visualize and solve complex problems, establish international relations, and embolden (even reignite) democracy. More than anything, what this lens offers us is a limitless capacity to handle all that comes our way, no matter how complex or unanticipated.

We are on the cusp of this renaissance, not in the midst of it. For all that science has contributed to our lives in the past half century, it hasn't yet universally changed the way we think. And it won't unless we understand and address why.

In times of uncertainty, fear becomes the overwhelming faculty. People seek out quick and easy answers and comfort, and religion offers both. It is soothing to think that when the ground shakes—killing two hundred thousand people in an already destitute nation—there is a grander purpose to it all, an underlying beneficent plan. Much less comforting is the objective explanation: that the Earth's tectonic plates constantly move and grind against one another, so there is compression and subduction, and the occasional violent rupture—a terrifying, but ultimately random and meaningless event. But it is only by studying those geological forces—and today we can—and embracing the findings, that we will be able to better predict earthquakes and save lives. We're not there yet . . .

Because science and religion are at war, embroiled in a battle that has not strengthened either side; it has served only to strengthen the bases, not convert the masses. (And the base for religion is a lot bigger than the base for science!) That's not to say the "culture war" has not been an intellectually useful exercise; it has. But religion will not overturn science and science will not overturn religion; both are too fundamentally rooted in society—the global economy is built on science; more than two-thirds of the world's population believes in a God. Pursuing each other's demise is ultimately deleterious to both and compromises the much more important goal of actually improving the state of the world. For the benefit of humanity, we need most of the seven billion people on this planet to embrace science as the way forward—we need them to use vaccines, fund stem cell labs, and support carbon taxes, not march against them.

The trouble with (commonly practiced organized) religion is not God. It is that it offers truth without a way of questioning it. Worse, it punishes you for questioning it. We simply cannot move forward in these times with this dogma.

Every child around the world should be taught to embrace the scientific method (this is what scientific literacy must mean in this century)—before they hear about God. And we should be reminded always that we have the right to question everything—that changing our minds with new evidence is a virtue not a cause for condemnation. If one then chooses faith and subjects it to permanent scrutiny, then religion is no longer the enemy. In the interest of global stability, science must precede faith but not seek to overturn it.

I should note that one of the practical problems here is that, unlike religion, science has no ordained leaders with the pulpit to make this plea. Science does not issue talking points. Ironically, this weakness, this absence of idols and enforced consensus, confers unique strength and stability to science. So perhaps instead, the world's priests and sheikhs and rabbis might take up the position.

Richard Feynman, the Nobel Prize–winning physicist, once told this story on television:

> I have a friend who's an artist and he's sometimes taken a view which I don't agree with very well. He'll hold up a flower and say, "look how beautiful it is," and I'll agree. And he says, "you see, I, as an artist, can see how beautiful this is, but you as a scientist, oh, take this all apart and it becomes a dull thing." And I think he's kind of nutty. First of all, the beauty that he sees is available to other people and to me too, although I might not be quite as refined aesthetically as he is.

But I can appreciate the beauty of a flower. At the same time, I see much more about the flower than he sees. I could imagine the cells in there, the complicated actions inside which also have a beauty. . . . The fact that the colors in the flower are evolved in order to attract insects to pollinate it is interesting—it means that insects can see the color. It adds a question—does this aesthetic sense also exist in the lower forms? Why is it aesthetic? . . . A science knowledge only adds to the excitement and mystery and the awe of a flower. It only adds. I don't understand how it subtracts.

It is one thing to accept science; it is another to embrace it. For some, the path to scientific thinking will be perfunctory. For most, though, it will need to be beautiful, inspiring, and fulfilling at the same time.

Millions of Soviet children cheered when Laika became the first living creature to orbit the Earth in 1957. Millions of American kids listened in rapt silence when Neil Armstrong, touching down on the lunar surface, delivered his indelible "One step" communiqué. An estimated six hundred million people worldwide—many of them children—watched the scene unfold on television: two grainy white shapes bobbing across the rocky moonscape, the first human beings to touch extraterrestrial soil. Two years earlier, when Robert Wilson, the director of Fermilab, was called to justify the multimillion-dollar particle accelerator to Congress, Wilson argued:

It has only to do with the respect with which we regard one another, the dignity of men, our love of culture. It has to do with: Are we good painters, good sculptors, great poets? I mean all the things we really venerate in our country and are patriotic about. It has nothing to do directly with defending our country except to make it worth defending.

We should all see science the way scientists do.

To do advanced physics at Fermilab or CERN takes advanced degrees. And we cannot all study dark matter. But that should not imply that we must remain on the receiving end of science rather than engaging with it as participants. We can all use the methodology, we can all contribute knowledge, and today, we can even all participate in some of the process.

Consider this: You can cook without mastering Escoffier. You can close your eyes and pray without reading Arabic. You can be an artist with just a paintbrush. The perception that the barrier to entry for science is somehow higher is wrong. It is probably even easier and more intuitive to think scientifically than it is to draw well or serve an ace . . . because we were once all scientists. It's how we thought as children, when we looked out at the ocean and wondered, "Why is it blue?" When we followed that question with further questions, and then with our own experiments: a jarful of salty water carried home from the beach answered: "Is it blue everywhere?" Creeping down to the waterfront well after dusk clinched the response to: "Does it stay blue at night?" If you wanted to understand how a butterfly flies, you examined it. You notice it has wings—two forewings and two hindwings, each pair attached to a different part of its body. An ant doesn't have wings, and a beetle doesn't appear to have any either . . . but there they are, hidden under a hard-shelled coat. Then, if you want to learn more, you crack open a biology book. You start adding information, observing more carefully, making richer hypotheses, and conducting experiments—and you do it because you want to, because it starts to become extremely gratifying. And then you start to understand something, and you begin making connections, seeing relationships and dependencies, and applying that knowledge . . . and a traffic circle suddenly resembles a horde of food-gathering ants.

The burgeoning "citizen science" movement coupled with the rise of mobile devices and social media allows anyone to participate in the process of science—counting migrating birds, collecting water samples, processing the search for extraterrestrial intelligence. . . . And the deepening relationship between science and design is opening up new means of physical and emotional interaction with the revolutions that science spurs.

Science is necessarily flat and open. Any idea can be overturned at anytime by anyone. This is fundamental to the culture and stability of science. Although they must ultimately hold up to the other tenets of science—reproducibility and falsifiability among them—the best ideas can come from anywhere. And they do. They come from architects conceiving higher dimensional spaces, from game designers toying with astrobiology, and from novelists thinking about memory.

Science Is Culture is a book of ideas from the leading edge of this revolution. It is the result of five years of conversations that we curated at *Seed* for a department called the Seed Salon, bringing together a community of artists and physicists, writers and designers, architects and geneticists . . . to explore themes of common interest to us all: the basis of morality, the nature of truth, the fundamental limits of knowledge . . .

This is a book about the renaissance on deck.

Adam Bly
June 2010

One

Evolutionary Philosophy
Edward O. Wilson and Daniel C. Dennett

The biologist and the philosopher meet up to talk about evolution, ethics, and the origins of religion.

Edward O. Wilson is a biologist. Daniel C. Dennett is a philosopher. Both believe that understanding evolution is essential to understanding our humanity. Despite an incipient blizzard, they met up to talk about God, evolution, incest, social norms, and, of course, ants. This conversation took place in March 2004, in Edward O. Wilson's office at the Museum of Comparative Zoology at Harvard.

Edward O. Wilson is a distinguished biologist, naturalist, and author whose seminal works—including *Sociobiology, The Diversity of Life, Consilience,* and the Pulitzer Prize winners *The Ants* and *On Human Nature*—have led to insights across a wide range of fields, from evolution to animal and human behavior. As the intellectual force behind the Encyclopedia of Life, a free, online collaborative portal that will document all living species known to science, Wilson continues his work as educator, researcher, and champion of biodiversity conservation. He is currently University Research Professor Emeritus and Honorary Curator in Entomology at Harvard University.

Daniel C. Dennett is a philosopher whose research centers on models of human consciousness, the conceptual foundations of evolutionary biology, and the ideas behind the practice of science itself. Though Dennett has gained a high profile as one of the "new atheists," he is perhaps best known in academic circles for his "multiple drafts" (or "fame in the brain") model of human consciousness, which describes a computational architecture that could explain the stream of consciousness. Author of numerous books, including *Content and Consciousness, Elbow Room, Darwin's Dangerous Idea,* and *Breaking the Spell,* Dennett is currently the co-director of the Center for Cognitive Studies, the Austin B. Fletcher Professor of Philosophy, and a university professor at Tufts University.

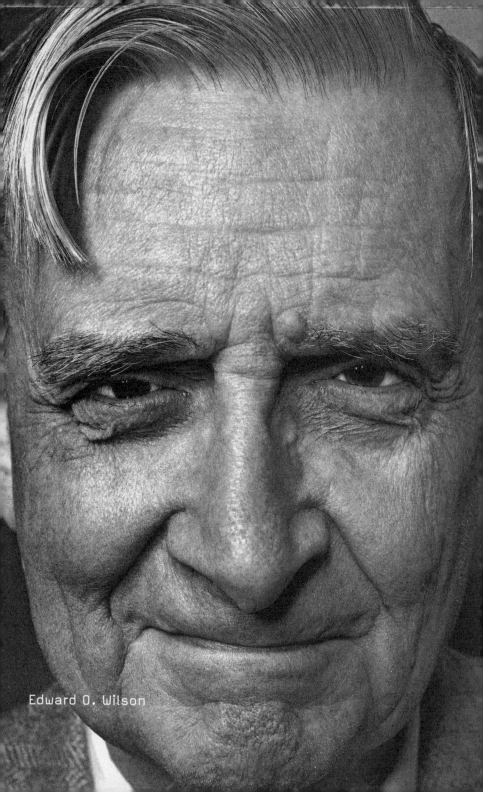
Edward O. Wilson

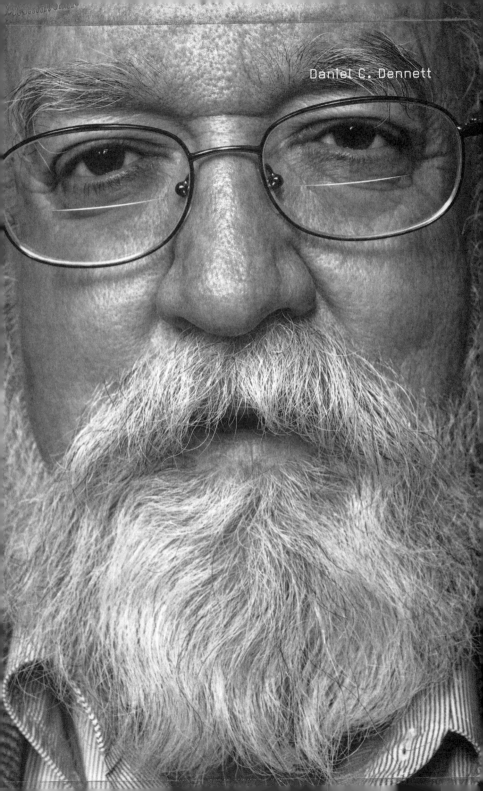
Daniel C. Dennett

Edward O. Wilson: It seems to me that the thing we have in common and why we can talk together—whereas I wouldn't be able to talk to most people in the humanities—is the perception that evolution is the key to understanding the human species. Where, in your judgment, is philosophy going, particularly the philosophy of science? Is there any merit to what Bertrand Russell once said, that science is what we know and philosophy is what we don't know? How do you see the picture between philosophy and science?

Daniel C. Dennett: I think that philosophy of science has actually become much better. We had our early period, which was very programmatic and was quite prepared to just figure out where science should go and where it shouldn't. And that was brave and foolhardy and it had some great moments, but people saw better, and now people in the philosophy of science have to be well trained. You quoted Bertrand Russell. Science is what we know; philosophy is what we don't know. I'm actually content with that, because I view philosophy as what you're doing when you don't know what the right questions are. And that's not trivial. If you can help sort out the bafflement and controversy and smoke and battle, that's work worth doing.

You may want to disagree with this, but it probably isn't too important for an entomologist to know the history of the field going back to the eighteenth century. But it really is important to know the history of philosophy if you're going to do philosophy, and the reason is actually very simple. The history of philosophy is a history of very tempting mistakes, and the people that we study in the history of philosophy—Plato and Aristotle and Kant and all the rest—they were not dummies. They were really smart people and they made stunning errors. These are very tempting mistakes. So you really have to learn the history of philosophy if you're going to do it well. Or you have to learn some of it. Because otherwise you just reinvent the wheel. You end up falling in the same old traps.

EW: Well put. I tried to encapsulate that similar notion by saying that philosophy consists of the history of failed models of the brain.

DD: Not just failed models of the brain, failed models of everything! [*Laughs*]

EW: [*Laughs*] It's true. Let me press this matter just a little bit further without descending into the abyss. Suppose that a young neuroscientist with a conception of how the brain might work proceeded with the most advanced technology. Suppose he chose not to look at the blind alleys of the past but to press on from this new phenomenology that's being produced. Wouldn't he succeed just as well?

DD: No, I don't think so. I mean, it's possible. But I think the odds against it are actually very high. It's not as if he would be proceeding without philosophy. It's just that he would be proceeding with seat-of-the-pants philosophy—with his gut instincts and his homegrown intuitions and what seems to make sense to him. Now, if you're lucky, that carries you on just fine. But on this topic in particular, on the mind, we can be pretty sure that some things we take for granted are just wrong. If they were intuitive, we would have gotten them a long time ago. The chances of just grubbing your way through the pitfalls and making it to the finish line with really good theory, all I can say is, "Go ahead and try, buddy, but I don't think your chances of success are very high."

EW: If you have a map of the minefield, use it.

DD: Right. Take advantage.

EW: It doesn't hurt to take Philosophy of Science 101.

DD: Yes.

EW: You know, you certainly cannot proceed very far on any of the great questions of philosophy, including ethics and the significance of religion, without an understanding of the full history of humanity. And by that I mean the deep history going back millions of years.

DD: Exactly.

EW: It is when consideration of the great questions of the human existence are limited, as they typically are, to the last ten thousand years, then you really are only considering a thin slice of time for human history. To many scholars, that might seem like all you need, but if you believe that the brain is not a blank slate, that there are deeply embedded programs of prepared learning that guide people in their mental development, then it makes sense to try to understand the deep history of humanity.

DD: Absolutely. I think that the great thing about Darwin's idea is that it's a unifying idea. It is the one theoretical perspective that has the resources, not yet fully exercised, to unite the worlds of meaning and beauty and truth and human aspirations with the world of matter and motion. As you say, the last ten thousand years are a twinkling of an eye. And if you don't understand how this is a very recent twinkling—a series of magnificent effects which had to be the result of things laid down eons before—then you're going to misconstrue your own treasures.

But beyond that it requires recognizing that such a thing as human creativity is itself a fundamentally mechanical and algorithmic process that's fed by many streams. Every single stream, every source of meaning in the world, ultimately has to be a fruit of the tree of life, and that means you have to ground it ultimately in evolution. But the idea that evolution can tie these two great aspects of the universe together is very unsettling to many people.

EW: We've agreed in our writings, both of us, that the key question to answer is exactly how the creative process came to be and why it's so extremely rare. In fact it is singular in three and a half billion years of evolution.

DD: Yes.

EW: I've dealt with this question in recent research and writings on the origin of the social insects and the fact of animal society generally, in which there's a high degree of altruism and the development of a worker caste. This is a conundrum that Darwin himself faced and thought would be fatal to his theory if he could not explain how you evolve worker ants that cannot reproduce themselves by natural selection. Of course, he essentially resolved it. The point I wanted to raise, however, in connection with this, is not how to resolve it, but rather to point out how extremely rare it is.

I recently made a count of all known cases of animal societies, in which there's a reproductive caste and there's a sterile worker caste. And the number is about twenty. And that is out of tens of thousands of independent evolving lines.

DD: I didn't know there were that many aside from the famous naked mole rats and us.

EW: Well, the rest of them are outside of that one mammal case. Or in the arthropods. In shrimp, aphids, beetles, and so on. But this is an extremely high bar to jump. So much so that even though evolving creatures in the land and sea have been poised to cross this bar, it's only been done, so far as we know, twenty times. And this takes us back to the question of why evolution has produced only one intelligent, self-reflective species like ourselves. I don't think you and I can sit here and answer that question.

DD: I don't.

EW: But it certainly has got to be one of the great remaining problems of science.

DD: But I also think there's a trap if we ask it in a certain way. I don't think you're going to fall in that trap, but a lot of people would. Evolutionary theory is unique in all of science in that it's a theory of things that almost never happen. Every birth in every lineage is a *potential* speciation event, but almost none of them are. The whole biosphere depends on these things that almost never happen. Mutations are almost never good. But it's the ones that are advantageous that do all the work. So it's tempting to ask a question like the one that you've just asked about why sociality doesn't emerge more often. Well, I don't think there's a reason why more of them don't, because like everything else in evolution, amazing things result, and one should simply get used to the fact that you don't have to explain why it doesn't happen. You only have to show the sufficient conditions, and every now and then they arise.

EW: We're back to the absolute necessity of understanding evolution, its causes and the many roads it's taken, in order to understand the human condition.

DD: Yes.

EW: I like to look at phylogeny, the actual step-by-step of evolution as it's occurring, as a maze. Species enter at one point in a certain condition. Typically we call this a primitive condition and of course it can be the opposite. It can be they're entering in order to go downhill and become parasites or cave dwellers or something regressive.

DD: Tapeworms, for instance.

EW: But let us say that it's a species that enters the niche and starts to evolve and divide into different species. What they have entered is a great maze of possibilities. A maze including the other species that serves as competitors, as biosphere engineers, as predators and food, which is constantly shifting. Most of those species passing through that maze have the opportunity to change from a purely solitary lifestyle to an advanced social lifestyle with division of labor between reproductive individuals and workers, and helpers that contribute their labor without reproducing. Most species will not make it because they will go up blind alleys. But a very small number will make their way through and come close to the breakout point. We find in those wasp and cockroach species that broke through and created what actually amounts to three-quarters of the biomass of all insects.

DD: Really?

EW: That's right. At least in a tropical environment. It's just an amazing success these social insects had. We find that the lines that achieved it evidently were pre-adapted, meaning they had reached a point where they had everything they needed. The right kind of food, the right kind of advantage that comes from living in a group, the right kind of environment that offers a stable environment.

DD: So we understand ants, termites, and other social insects by retracing the steps they've taken through the maze, and it seems to me that we need to do the same thing with explaining this great unique human breakthrough.

EW: You come to this question that you intimated earlier, having to do with our current lack of a true understanding of the human mind. That is the still hard-drawn distinction made between the great branches of learning and the belief of many on the social sciences and humanities side that there's an indelible barrier—

DD: An unbridgeable chasm.

EW: And that it's pointless and misleading and maybe even harmful to speak of it. However, I see this not as a chasm but a large broad area of mostly unexplored terrain. Certainly the mind has a central site to explore there.

DD: It's like the two ships in the sea trying to tie alongside each other. We throw a whole bunch of lines across, but there's gonna be a lot of bumping and shoving and people pulling too hard and that's what we're in right now. There's a lot of legitimate anxiety as these two domains have been so independent of each other for so long. Once we get them tied together tight, it'll be fine, but in the meantime there's gonna be an awful lot of jostling and bumping, and that's what we're going through, I think.

EW: There's a lot of anxiety on the part of humanities scholars. I think it's in part because right now natural scientists have the greatest amount of money and prestige. But quite far from damaging the humanities by coming closer, collaborating and building the bridge, or, as you put it, tying the boats together, is one way to gain in strength and prestige.

DD: I think so.

EW: They should look at it as a process of cooperation and claim that middle ground, a good part of it, for themselves. I have noticed that quite a few young scholars in literary theory, and even the visual arts, are beginning to explore this possibility seriously. They're up against a lot of tenured old guard, but nonetheless they see it, and I think correctly, as their chance to do something really new.

DD: But at the same time there're people pulling to some degree in the wrong direction there, too. To my way of thinking, the real problem right now is that we have enthusiasts who oversell, and then we have to go in and correct the overselling and say wait a minute, that's a little too swift. And if we don't do that, then we just seem to confirm for people in the humanities who are most skeptical that this isn't an honestly brokered scientific investigation, but rather an ideological takeover.

EW: It is particularly difficult in the case of ethics, because the normative [refers to guidelines or norms; prescriptive] and the objectively scientific are so closely mixed. They're talking about developing a science of normative reasoning. That's why I think that the evolutionary approach is absolutely critical.

DD: It is, but here is perhaps the one area where you and I have some disagreements. I think *Consilience* is a magnificent book, but there are some places where I've got to differ with you, where I think you've overlooked some points coming from the other side. The place where you draw the line between normativity and factual exploration is not where I would draw the line, and so this is an area where we should sort out the few remaining points of difference.

Think about some more trivial domain of normativity, like how to play good chess. Now, what is the status of the principles of good chess play? Not the rules, but the principles of good, winning chess play. It's normative. Now, in this trivial case, we know what the goal is. The goal is checkmate, although we can imagine other goals, too. We can imagine somebody writing a book called how to play

Edward O. Wilson and Daniel C. Dennett

polite, winning chess, where there would be a particular prohibition of being sort of brutal, the way, say, Bobby Fischer often was in his games. You know, don't humiliate your opponent. Well, that's a different set of norms.

EW: Aren't you conflating game theory with ethics?

DD: Am I conflating game theory with ethics? In a way I am, and on purpose. Game theory is a particular instance, a particular species of hypothetical imperatives. It's looking at two-person or many-person games, and saying this is what the optimal strategy is.

EW: Yes. To achieve a certain goal. But the goal that you pick in playing a game of chess depends upon ethics. That is, you just mentioned the thought: "Shall I humiliate and therefore make the game more exciting to the crowd?" which is one goal. And that is an ethical choice, it seems to me. Particularly since you're going to be dealing with the feelings and the mood of the group. Or one could ask, "Shall I just do it completely mechanically," or, you know, "hold the bum up until the ninth round?"

So that seems to me to raise the question of how to deal with normative issues, and that is why I believe that ethics, or, shall I say morality, is the behavior by which, if I've got it right, we effect a common ethical goal or program. I believe that the evolutionary approach is necessary to disentangle normative from objective fact-seeking. If the naturalistic view of the origin of ethics is correct, that's essentially an evolutionary view.

DD: Yeah.

EW: Then ethics, the code that we develop, the set of goals that define a culture, evolves to different forms in different societies. Sparta was different from Athens. A society that has come to sacralize its rules and codes of behavior by fundamentalist religion is very different from a more secular country. Iran is different from France. And tracing how their ethics arose and evolved with time can be an objective study.

DD: Indeed it can. But suppose we've done that study. In the same spirit we can say, well now, some people play go and some people play chess and some people used to play chess without castling and we *could* play chess where you move the king two moves instead of one. Which game should we play? Now, that's not an insane question, and it's not just a historical question. It's asking what game should we want to play. And we can compare these different games and decide which ones best serve our purposes. Now the process of doing that can be done well or ill. If it's done well, we can use the very investigation of which game we should play to clarify our own thinking on what game it would be best to play. The question, "What game would it be best to play?" is not an idle question. And that is the ultimate ethical question, and it's a normative question. You don't answer that question, you clarify, you illuminate it, but you don't answer that question by studying the evolutionary history.

EW: You talk like a philosopher on this issue, Dan, and that is in abstractions.

DD: That's what I am.

EW: Okay. [*Laughs*] I see your point. But let's talk about some of these enormously important cultural differences that are now edge issues in the United States, and see how an evolutionary approach could help to resolve them. Take homosexuality. And perhaps in your philosopher's all-comprehensive mind, I'm moving too quickly to example, but—

DD: No, not at all.

EW: Okay. Let me suggest homosexuality. Fundamentally at the base of arguing trivialities such as whether marriage is allowed is something that's fixed in the minds of people: whether it is natural. Catholic theologians would call it obedient to natural law. One seeks in vain—at least I seek in vain—to find out exactly what the Church means by natural law. But put that aside. People either believe that it is natural in a way that is harmless to society or maybe even to some extent good, whereas others believe it is unnatural and dissolutive in terms of its effects on society. Well, how could we resolve that? We seldom see in op-ed pages and editorials the question of naturalness of homosexuality. And there is a lot to be said: that it occurs at a fairly constant frequency across societies; that it does have at least a partial biological basis; that there is some evidence from ethnographies of other, simpler societies, that homosexuals have a positive value in the survival of families and tribes to which they belong. If these issues were pursued with reference to the key question of what is natural, and then we get down to the next level below that, what is natural in terms of natural selection and human history, we might, if not resolve the difference, at least shrink it somewhat.

DD: Yes. But I think you would agree that part of the problem with setting the question up this way is that *natural* means different things to different people. Let's make a list of some other things that are natural. Shortsightedness is natural. Hypertension is natural. Obesity is natural if you eat too much. There are many things that have deep evolutionary roots that are natural. But one of the glories of civilizations is we've learned to adjust things that might be natural but we don't like them. So I think natural cuts two ways. Artificial fibers for many purposes are just fine. And for some purposes natural foods are no better than artificial foods. Nudists say it's natural to go without clothes, and yet there's a sense in which they're just wrong.

EW: I didn't mean to say that the guideline of ethical judgment is what is natural. Because of course tribal warfare is natural . . .

DD: Exactly.

EW: And has always been advantageous in terms of its survival value for groups. And there are many other human behaviors that were natural because they evolved in hunter-gatherer times and that are destructive now because they are so easily overdone. So this then brings us back to the question of the distinction between the ill effects of excess, which of course one has in overeating and the like, and something as fundamental as sexual preference. Sexual preference is not, so far as we know, the result of doing too much of one thing or another. It appears to have a real biological basis, and, if we've pegged it right, may have gotten as common as it is because in past times it was advantageous for families or tribes. So at the very least we're in a position here of trying to moderate people's behavior or develop the arts of diplomacy to reduce tension. We're in a position here of arguing whether we should have tolerance or not.

DD: First of all I want to say that I deplore the attitude of those religious groups and others who want to deny full rights to people who are homosexual. I just think the issue isn't really about whether it's in their genes or whether they decide it. I don't think it really makes any difference. I don't think we should make our policy with regard to homosexuality hostage to the question of whether or not we find that there is a specific genetic disposition for homosexuality. That's not what the moral decision should hinge on.

EW: I agree, but you see what we're doing here is not making the acceptance of homosexuality hostage to the idea of natural. We're preventing the other side from making homophobia hostage to their idea of what's natural.

DD: That's fine. I agree.

EW: Let's take another example. Incest avoidance. There have actually been arguments that incest taboos were just another form of sexual repression of free sexual behavior. But we now know in some detail that our natural avoidance of incest at the family level has a deep genetic algorithm behind it. And we know the reason for that algorithm without any doubt at all in terms of natural selection. Namely that first-order incest between people—in other words, people who are related by half their genes—is extraordinarily destructive in the production or the rate of production of abnormal children. Now, here we have arrived at something that both by the

Edward O. Wilson and Daniel C. Dennett

structures of religious belief in which it's all thought to be the desire of God, and by secular belief, which arrives by studies of human nature and the consequences of actions, are in agreement.

DD: Yes.

EW: So once all that information is put together, particularly the biological information, then the matter becomes a no-brainer for all sides. Well, I'm in the hope that with the evolutionary approach to ethical interpretation of ethical precepts we might narrow the gaps in thinking between different cultures, different groups within cultures.

DD: I think we can. I think it's always going to be indirect. (And you'll say, well, this is the philosopher speaking, because he wants everything to have many curlicues. He's not gonna go straight for the win.) But I do think that the important point is that there is a logical independence between the scientific facts and the normative questions that we all have to face and answer, ultimately politically. We have to reflect on it with the largest group of people and then we have to decide as a group what we think we ought to do. That's the only way we can solve that problem.

EW: That's certainly my position. I was talking about the best route using science.

DD: And I think that using science to inform that political process is really important, and we're in complete agreement there, too. But I want to resist any suggestion, which you may not be making, that the scientific discoveries in and of themselves settle any of these issues. I don't think they settle the issues. I think they only illuminate the issues. I think that we can't draw any conclusion at all about what we ought to do from even the most perfect knowledge of how we got to be what we are. That gives us a great factual basis then to sit down and say, well, okay, which aspects of this are we content with? We have evolved to have these dispositions and these features, and wouldn't it be good if we could do something about it. There's still room for those questions. It's still a legitimate inquiry.

EW: I think we've come back off our detour to the main highway in agreement. Certainly the precepts of ethics and the morality that comes from them is reached by consensus. It's not dictated from above and it's not dictated by our past evolutionary history. But certainly the past history is key to working out the agreement.

DD: Absolutely.

EW: I would be very interested before venturing any opinion I have, not to put you at a disadvantage, to know what is your feeling about the conflict between science and religion generally? And particularly the more traditional, organized religions.

DD: I'm currently working on a book on religion and science.

EW: I'm not surprised.

DD: And the point of the book is, and perhaps this is even comically philosophical, to acknowledge in a Socratic fashion that I just don't know. Because we don't know enough about religion. We haven't studied religion with the tools of science the way we should. So my first plea is to break the taboo against studying religion as a natural phenomenon. We have to study religion with the same intensity that we study global warming and El Niño. Religion as a phenomenon is one of the most important and influential phenomena in the world, and we are embarrassingly ignorant about it.

Several hundred years ago at the triumph of the Enlightenment, which you and I both admire and wish to restore, many wise, well-informed people were very sure that now that we had science and enlightenment upon us, religion would soon die out. They were colossally wrong. Religion is still thriving in our midst. So what didn't they understand? What is it about religion that makes it so important to so many people? And how did it evolve? There's an evolutionary answer there, too. And we don't know. There are some very interesting forays into this examination of what you can call the evolutionary biology of religion, and I am hot on the trail of them. Until we know more about how religion evolved to be what it is, we won't know the right way to deal with it. We've certainly made a lot of mistakes in the past. Suppression of religion is almost certainly a terrible idea. Worse than Prohibition. Until we realize why religion is so important to so many people we are powerless to have policies that are wise and well-informed.

EW: Struggle as I might to find a disagreement, I can't. I agree entirely. I think we already have a rough idea of the origin of the overpowering desire to acquire religious belief and all of them are fundamental. One is the great advantage there is to having a common belief system within a society. It can be used to unite it. It can be used to excite it, to infuse courage and persistence in the face of hardship. That's also true at the family level. It just has great survival power and there isn't much doubt in my mind that some part of the brain is hard-wired to acquire through prepared learning that type of allegiance and the strong emotions that go with it.

Of course, we also have a deep fear of death because we're the first species to understand our own mortality. Religion helps us to escape that. I believe that we will gradually come to understand religion. Unfortunately, there's still so much resistance, not just in Islam but also in Western cultures, that in this country you're likely to have a fatwa of sorts issued against you if you start probing too deeply.

DD: And this is precisely why I am writing this book addressing deeply religious people and trying to explain to them why they themselves should want to explore scientifically their own religion, their own religious beliefs. Because they are very sure that their religion is the most important thing in the world, and they may be right, but they don't know. And they owe it to the rest of us to explore the grounds for their confidence that they hold the moral high ground. And if they can show us that they do hold the moral high ground then I'll join forces with them in a moment. But I don't accept their claim that they "just know" that religion is the right path, and that I have no business asking how they know that. If they want to enlist those of us who are skeptical about where the moral high ground lies, then they have to be willing to demonstrate. They have to be willing to sit still while we take a good hard look to see if they're right. They may be right. They may not. But they themselves should be willing to undergo this.

My thought experiment to understand their position—and this is as much for my benefit as for anybody else's—is to imagine that some scientist comes along and says, "It turns out music's really bad for you. We should try to eliminate music from the world. We should forbid children to take music lessons and strictly limit our own listening time because it turns out that contrary to what we've thought for millennia, music is bad for your brain and bad for society." If somebody came along and said this, first of all I would feel a visceral urge to disagree with this person, because I'm not sure I would want to live in a world without music. Well, that's the way many people feel about religion. But then I reflect: If somebody came forward and said, "Look, are you so sure music is a good thing?" I would say, "Well, in fact, I'm so sure I'm willing to put it to the test. Study us music-lovers to your heart's content. I want you to explore every aspect of music and if you show me that music is a pernicious thing—there are, after all, people who believe that, the Taliban believe that, for instance—then I'm gonna have a hard decision, because maybe I'll have to give up music." Now, I want to put the same proposition to the deeply religious. They have to consider enough about their religion so that they can see whether their belief that they hold the moral high ground is justified.

EW: They—and even the religious fundamentalists—may be more willing to consider an argument like that, especially in view that they have now had demonstrated to them that the world is filled with people willing to commit suicide in the name of religion. Which bespeaks a deep belief in a different system of beliefs and moral guidelines.

Edward O. Wilson and Daniel C. Dennett

I know from personal experience, having been raised a Southern Baptist, that the quality of religion in billions of people is measured by the intensity of their belief and commitment, not by any factual information that they might be presented with. Also I would expect—I'll just make a broad prediction on my part—that we need the rites of passage. We need the great music and the great liturgies. We need the entire culture of religion and religion-like ceremonies and engagements and reaffirmations. This is so fundamental to human nature that it's like music.

DD: Yes.

EW: It doesn't necessarily carry with it, however, a specific mythology or an unwillingness to listen to the basic beliefs of people in other faiths. I'm inclined to think that religion will evolve as time goes on. In its constant contact with science and growing knowledge about where religion comes from, it will evolve to hollow out, in a way, its mythologies, and come to depend more on the manifest strengths of its ceremonies and reaffirmation of spirit.

DD: There's several radically different hypotheses about what's going to happen to religion in the future, and yours, I think, is one of the most hopeful. There are many, of course, who think that this benign evolution that you anticipate and hope for would be rendered less possible by the meddling intrusion of somebody asking the questions that I want to ask. They think, in fact, that I risk breaking the spell and making those very valuable ceremonial and allegiance phenomena less sustainable. Now I take that to be a really serious worry, and that's one that I want to explore most carefully in this book, because I certainly don't want to vandalize that aspect of religion. I want to explore the question of whether your hypothesis is reasonable and plausible, and if so, is this the direction in which we want to push. But in order to do that we're gonna still have to look pretty hard.

EW: I agree. That's the line of investigation we need to follow, and so I'm optimistic in part because I still am deeply moved by an evangelical hymn, or a Catholic high funeral Mass. In other words, the emotions stay deeply embedded there. The ability to feel them as deeply as a traditionally religious person feels them is not dependent on a particular mythology. It has to do with my personal experience. My personal experience makes me believe I'm right.

DD: Yeah. I agree. My personal experience is not science, okay. You know, I think we should be coming to a close.

DD: Yes.

EW: But I was just thinking, not trying to serve as a moderator-interrogator, but I'm just wondering if you or I haven't really got this clear. In my mind, I think we've covered a lot of ground.

DD: I think we've covered a lot of ground. I could talk with you for hours about *Consilience* and your original book, *Sociobiology*, and *On Human Nature*.

EW: Well, thank you for saying that. And I'm not just expressing politesse when I say you've thrown across some real hawsers yourself. And I'm thinking perhaps down the line, if we ever have more leisure time, it would be good to continue this conversation in a different setting.

DD: Sure.

EW: You know, maybe a seminar or something like that. And do it maybe in front of a group of students and engage them. That sounds to me like a perfect course to give.

DD: Would be fun, wouldn't it?

EW: Yes. But we can talk about that later. I don't know *whether* we'll ever have time. You've seen the way I live. I can hardly find enough time to go to lunch. At any rate, the point is that I think we covered a lot of ground and I just, do you have anything in the way of a wrap-up thing you want to say?

Edward O. Wilson and Daniel C. Dennett

DD: I think it's important that you and I don't agree completely, but when we discuss our disagreements, it's always constructive.

EW: Yes. Well the blankest part of the map is where naturally we're going to diverge, and I would expect that.

DD: Yes.

EW: In fact, if we didn't diverge on the blank parts of the map, we better reexamine just how dogmatic we have become.

DD: [*Laughs*] Exactly. Well, we've let it all hang out here.

EW: And now we definitely think we're done. I really enjoyed it.

DD: Yes. I knew I was going to enjoy this. ∞

Two

The Problems of Consciousness
Rebecca Goldstein and Steven Pinker

The psychologist and the novelist discuss storytelling, empathy, and human nature.

Steven Pinker is a psychologist. Rebecca Goldstein is a novelist. Both are obsessed with realism and the pursuit of objective knowledge. They met up in 2004 at Manhattan's Algonquin Hotel to talk about consciousness, storytelling, game theory, and gossip.

Steven Pinker is an experimental psychologist who has taught at Stanford, MIT, and Harvard, where he is currently the Harvard College Professor of Psychology. Pinker's research on language and cognition has won prizes from the National Academy of Sciences, the Royal Institution of Great Britain, the American Psychological Association, and the Cognitive *Neuroscience* Society. His books, including *The Language Instinct, How the Mind Works, Words and Rules, The Blank Slate*, and *The Stuff of Thought*, have earned numerous prizes, together with two short-listings for the Pulitzer Prize.

Philosopher Rebecca Goldstein is the author of six novels (*The Mind-Body Problem; The Late-Summer Passion of a Woman of Mind; The Dark Sister; Mazel; Properties of Light: A Novel of Love, Betrayal, and Quantum Physics;* and *36 Arguments for the Existence of God: A Work of Fiction*) and a collection of stories (*Strange Attractors*). She has also published to critical acclaim two works of nonfiction, exploring the works of Kurt Gödel and Baruch Spinoza. Among her honors are two Whiting Foundation Awards (one in philosophy, one in writing), two National Jewish Book Awards, the Edward Lewis Wallant Award, and the Prairie Schooner Best Short Story Award. In 1996 she was named a MacArthur Foundation Fellow, and since then she has been elected to the American Academy of Arts and Sciences, received Guggenheim and Radcliffe fellowships, and was designated a Humanist Laureate by the International Academy of Humanism.

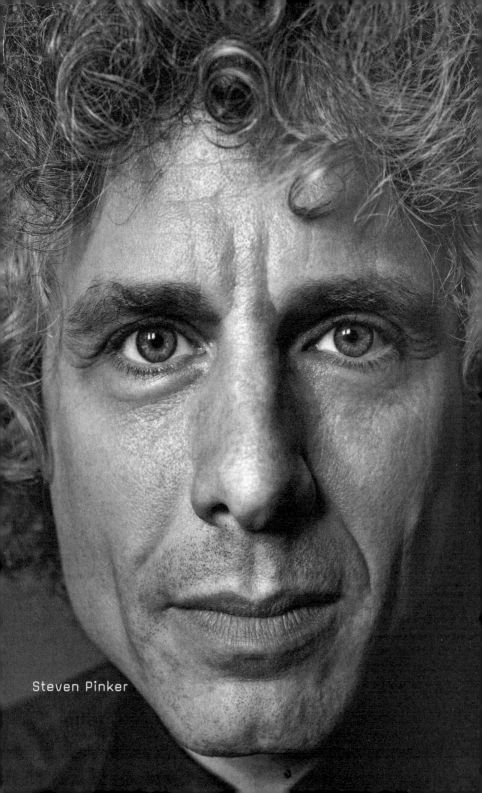
Steven Pinker

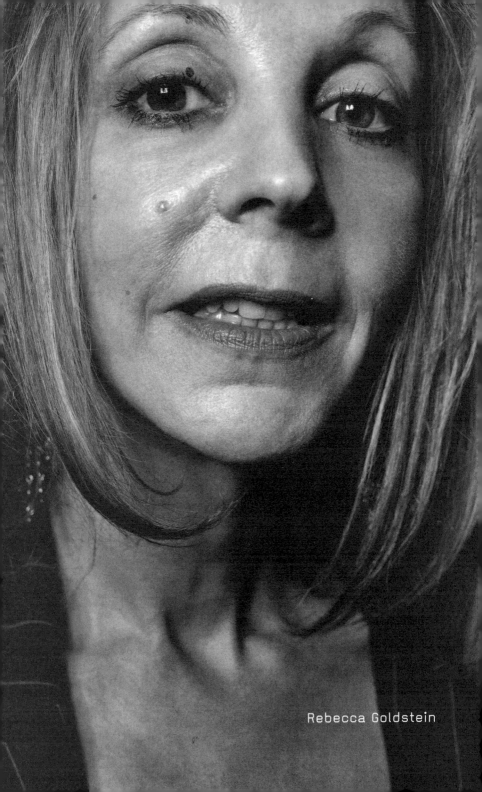

Rebecca Goldstein

Rebecca Goldstein: I've been a longtime fan of yours. I was blown away by *How the Mind Works.* So many cognitive faculties explained, and such a satisfying strategy for explanations: combining the computational theory of mind with evolutionary psychology. It's an elegant and powerful theory.

Steven Pinker: Thank you. I'm happy to return the compliment. Your first novel, *The Mind-Body Problem*, is a classic among people in my field, and I've enjoyed all your novels since. They've taught me a lot of philosophy and physics, and they're tremendously enjoyable as literary fictions.

It seems to me that you were ahead of your time in bringing themes from science into fiction. We're certainly seeing more of that now: Authors are incorporating ideas from science into their fiction—together with analytical philosophy, which I consider continuous with science. And scientists are seeking insight from literature and other cultural mediums.

Perhaps it's because many of the ideas that scientists and philosophers worry about are the same ideas that thoughtful people worry about. How can we know the truth? What is the relation between mind and body? What does it mean to do right or wrong? Is morality a product of our minds and tastes? Or does it somehow exist outside of us?

RG: These are questions you can hardly avoid if you're at all given to self-reflection, though the culture did try to avoid them for quite a few decades. Scientists were apt to call these big questions meaningless, and philosophers were even more likely to do the same.

SP: One question that seems to worry both of us regards realism—the idea that things really exist out there, and are not just social constructions or figments of our imagination. Your characters are obsessed with realism. *In Properties of Light*, the embittered physicist hates the standard interpretations of quantum mechanics in which particles don't really have physical properties until they are observed. And in *The Mind-Body Problem*, the mathematician Noam Himmel staunchly believed he was discovering aspects of reality.

RG: Yes. Noam's form of mathematical realism can strike laypeople as eccentric, but I was really just echoing the ideas of famous mathematicians, particularly G. H. Hardy's in *A Mathematician's Apology*. Say you ask a mathematician, "What do you think you're doing when you do mathematics? Is it like chess, where you make up some rules and see what you can or can't do with them? Is math just a bunch of empty tautologies? Or are mathematicians just discovering the implications of certain features of thinking, so that you're really in the business of psychology? Or are you mathematicians more like physicists, discovering the facts—not of the spatiotemporal world but of necessary truths about objective mathematical structures?" Most mathematicians would probably choose an option like the last.

SP: Cognitive scientists care about this question because they wonder whether math is just the exercise of certain faculties of an evolved mind—the number sense, spatial cognition, estimation. Of course this may not be incompatible with mathematical realism. Given a creature who, for good evolutionary reasons, has a mind that can grasp concepts like two, three, and addition, the nature of mathematical reality gives it no choice but to conclude that two plus three equals five.

But why would a novelist care so much about realism?

RG: The search for objective knowledge strikes me as a form of heroism. It can be deeply marked by egotism, of course, and great thinkers certainly aren't necessarily saints. Still, to subjugate yourself to the objective truth is a humbling experience. And the life of the mind is filled with passion, so it's a fit subject for art, as more novelists and playwrights and moviemakers are finding. Dramatizing the passion of knowledge-seeking is a good corrective for postmodern cynicism.

SP: One of your characters, Raizel Kaidish, was passionate about the idea that morality has an objective reality. Many people who believe in the reality of rocks and even numbers have particular trouble swallowing moral realism.

RG: They do and they don't. People are often committed to moral realism without realizing it. My students often argue for moral relativism on the grounds that anything else is intolerant of other points of view. They're not willing to regard tolerance relativistically. Tolerance is a moral value on the basis of which they inconsistently argue that there are no moral values.

SP: People are always realists when it comes to their own convictions.

RG: One of the arguments for realism—and I'm thinking here of an argument that I think you make about our realism in regard to the external world of three-dimensional objects—is to show that these beliefs are structural features of our thinking and we can't get along without them. When we deny them, we end up contradicting ourselves, and the impossibility of consistent doubt provides some evidence for their truth. These deep instincts for realism—whether for physical objects or moral values—may have evolved because these actually are realities, so it's useful to believe that they are.

SP: The biological argument against cultural relativism is that it is not just a set of social conventions that vary from culture to culture, like driving on the left or right, but has roots in human nature. But many people are just as nervous about this possibility, because they fear that it leaves morality as nothing more than a figment of the wiring of our brain, and still ultimately a sham.

But Raizel made me think that there really is a sense in which morality has a logic that exists outside of us, and that our moral sense may have evolved to grasp this moral reality in the same way that our faculty of depth perception evolved to deal with a world that really is three-dimensional.

RG: There is something in grasping another person in the full complexity of their own personhood that entails another domain of facts, facts about rights and obligations. There are ways that you can and can't morally treat a subject of experience, especially if that subject is a person. So if we're realists about other persons, perhaps we're forced to be moral realists. That's basically the Kantian argument.

SP: As long as your welfare depends on the actions of another person—as long as you're not a galactic overlord—you can't insist that they adhere to a code of behavior that you're not willing to adhere to yourself. Logically, you can't make the argument that you have privileged interests that other people lack—that there's something uniquely special about you in the universe.

RG: In the story you refer to, it's Raizel's mother, a Holocaust survivor, who makes an argument like the one you're making—that the immoralist is committing a sort of logical error in refusing to universalize the "mattering" he confers on his own person to other persons. Sometimes that mistake resides—as in the Holocaust—in not acknowledging the personhood of certain groups of people.

SP: Morality, at heart, is the idea that one's own perspective is not privileged—that the only coherent code of behavior takes a disinterested perspective that applies equally to oneself and to others.

RG: One of the many things I've learned from your work is that this kind of argument has its basis in our intuitive psychology. Part of the innate structure of our mind is a belief that other people are people in just the way we are. We need this belief to make sense of other people's actions and relate those actions to life. But there's a moral element in our acknowledging that other people have all the aspects of subjecthood that we do.

SP: The ability to take someone else's perspective may tie into the moral sense and the forces that led it to evolve—namely, the logic of reciprocity. If we live in a world in which each of us, in the fullness of time, will be in a position to do the other a favor—or at least refrain from hurting each other—and if we both end up better off if we help each other than if we hurt each other, then certain moral emotions are expected to evolve. For instance, sympathy, where we extend help to someone who we feel needs it, or gratitude, where we feel warm toward someone who extends help to us, or righteous anger, when we want to punish someone who withholds help or causes harm. So an aspect of reality—the inherent benefits of interchanging perspectives—may have shaped an aspect of the mind, namely the moral sense.

RG: I'm sympathetic to this account of moral reasoning, up to a point. But I don't feel that it provides the whole story of what it is to think morally, or the complete answer to why we all more or less naturally think morally, so that even the moral relativists in my classroom can't help slipping into moral realism. There's something very immediate, though also very complicated, in imagining yourself into another person that we don't get to by way of game-theoretic calculations. I think the rest of the moral story is tied up with a different kind of thinking—narrative thinking.

SP: And that brings us back to fiction. One problem for anyone like me who believes in a fixed human nature, including a fixed moral sense, is to explain how human behavior could have changed so radically over a few centuries or millennia. Much of the world has seen an end to slavery, to genocide for convenience, to torture as a routine form of criminal punishment, to capital punishment for property crimes, to human sacrifice, to rape as the spoils of war, to the ownership of women. We seem to be turning into a nicer species.

RG: There certainly are places on the globe right now where we're regressing dreadfully.

SP: But taking a millennium by millennium view, the twentieth century may . . .

RG: Seem like a bloodletting of horrible proportions, when in fact, statistically . . .

SP: We have a much lower rate of death in warfare than pre-state, hunter-gatherer societies.

RG: So we're getting less cruel.

SP: We are getting less cruel, and the question is, How? The philosopher Peter Singer offers a clue when he notes that there really does seem to be a universal capacity for empathy, but that by default people apply it only within the narrow circle of the family or village or clan. Over the millennia, the moral circle has expanded to encompass other clans, other tribes, and other races. The question is, Why did it happen? What stretched our innate capacity for empathy? And one answer is mediums that force us to take other people's perspectives, such as journalism, history, and realistic fiction.

RG: Storytelling does it.

SP: By allowing you to project yourself into the lives of people of different times and places and races, in a way that wouldn't spontaneously occur to you, fiction can force you into the perspective of a person unlike yourself, who might otherwise seem subhuman.

RG: There's a fundamental role that storytelling is always playing in the moral life. To try to see somebody on their own terms, which is part of what it is to be moral, is to try to make sense of their world, to try to tell the story of their life as they would tell it. So in our real life, just in making sense of people's actions and in seeing them in the moral light, we're involved in storytelling.

SP: So you agree that fiction can expand a person's moral circle?

RG: I think storytelling, in general, has a moral use. To be in the throes of a story, to have one's emotions provoked by another's story, is not quite ethics, but it's kind of the shadow life of ethics. We train children by telling them stories. We try to get them to feel their way into other lives, and that itself is something. If we had no capacity for that there would be no hope. It would just be all rules that you would follow for fear of being punished if you didn't. And that doesn't amount to becoming a moral agent. But then storytelling can also correct that second sort of moral mistake, the one of not recognizing the essential personhood of certain groups outside your chosen sphere. And that's really the point you're addressing in speaking about the expansion of a person's moral circle and the role that fiction can play.

SP: One famous example, I guess, is *Uncle Tom's Cabin*.

RG: Lincoln, when he met Harriet Beecher Stowe, said, "So you're the little woman who wrote the book that made this great war!" He meant that she, as a novelist, got people to see the enslaved as people. And once you do see that, certain kinds of behavior became impossible.

I have a personal story about this. My sister was teaching English at an all-girl high school that was rather racist. So she brought in a short story, and she had the girls read it aloud. It was about a black man, and because the girls were so insensitive to this issue, they didn't get the clues about the identity of the protagonist until very late in the story, after they were hooked. She said it was like watching a moral awakening. They had been weeping over this man's story, inhabiting the point of view—would they ever feel the same again?

SP: That's a wonderful example of an answer to the question, How can a part of the mind that was there all along—the capacity for empathy, in this case—suddenly be extended to a new target, which may not be evolutionarily natural? Fiction can be a kind of moral technology.

RG: Storytelling is something that can awaken attentiveness, engagement, and empathy to a life that isn't one's own. And to be attentive, engaged, empathetic: that is moral.

SP: Of course, we don't only listen to stories to expand our moral sense. A puzzle that I wonder about is why our species takes so much pleasure in fictional narratives in general. Storytelling is universal; it's done in all cultures. And it emerges early in our lives, as we see from the delight that children take in stories. But why do people devote so much brain power to creating and appreciating tales of things that aren't true? Literature is a pack of lies. There never was a Hamlet. Eliza Doolittle never existed. But still, we can't get enough of it. And we also have the sense that storytelling is inherently worthy. It's not a waste of time to appreciate good fiction.

RG: So I'm going to ask the sort of question I learned to ask from studying you: How could all this storytelling have adaptive value? What good does it do the species to sit around enchanting itself with make-believe? And it really is a kind of enchantment. A writer feels it very much, because the enchantment of writing is of even greater intensity than being an enchanted reader. Plato said it was a kind of madness, which is why he didn't think very much of the literary writers. He exiled them from his utopia.

SP: You have delusions.

RG: You're spellbound. You hear voices. You're inhabiting another world. It's so thick, it's so deep, it's so profound, that it can seem like a half step away from madness. Except one is in control, and when the book is finished, the voices go away, which is always reassuring.

There's also the fact that when you are in that thick enchantment, so far away from reality, you're so much smarter, and that seems mystifying to me. Think of what Shakespeare knew about human nature, things you scientists can explain only years and years later. And the great writers still know more about human nature than you scientists do. There's something about this state of fictive enchantment that puts the enchanted writer in the way of very large truths, and that's an aspect that's very mysterious to me. And I'd like you to explain it.

SP: [Laughs] I don't know if I can do that, but I certainly have worried about the puzzle, too. Why is fiction so compelling? Part of the answer might be that it is a way in which we press our own pleasure buttons. There are good adaptive reasons for people to enjoy gossip, namely that it provides compromising information about other people, which allows us to do the equivalent of insider trading. In a sense, fiction is simulated gossip. You're a witness to the secret foibles of other people. They just happen not to exist.

RG: And it's true that the great themes of gossip—sex and violence—are the great themes of fiction.

SP: I've argued that the pleasure-button theory helps explain many art forms, like painting and music. They would be evolutionary by-products, not adaptations. But when it comes to fictional narratives, I suspect there is an adaptive benefit as well. One problem we all face is how to act in a world that presents a vast combinatorial space of possibilities, especially when it comes to other people. I can do any of ten things, and you can do any of ten things in response to each of those ten things, and I can do ten things in response to your response, and so on. There is an explosion of possibilities that no mortal mind can deduce in advance. What fiction might do is allow people to play out, in their mind's eye, hypothetical courses of action in hypothetical circumstances, which would then allow them to anticipate what would happen if they ever faced those situations in reality.

RG: If I'm a bored young wife, living with a provincial doctor, what will happen if I have a series of affairs with men above my social station and get into a great deal of debt that I can't tell my husband about? Flaubert figures out the consequences for us, so we don't have to go and actually live through Emma Bovary's bad decisions.

SP: Exactly. He answers the question, What's the worst that can happen?

For an adaptive hypothesis to be taken seriously, there should be some independent reason to believe that the trait really is a good engineering solution to the problem. We don't want to just invent any old story for why some part of the mind is useful, just because we know that the mind has that part. In this case, I think there is an independent rationale. It comes from the approach to artificial intelligence called case-based reasoning, in which the best way to solve a problem is to analogize it to some similar problem encountered in the past, rather than cranking out a set of deductions using logical rules. So the system keeps a library of cases in memory and refers back to them when solving a similar problem.

Perhaps fiction is a kind of case-based reasoning. It multiplies the number of scenarios that you have tucked away in your mind and that you can call on as a guide for a future action. Of course, for that to work, there have to be constraints in the fictional worlds. It can't be true that anything can happen, or else what plays out in a fictional plot would have no lessons for real life. And this gets us back to your observation that novelists feel they're at their most intelligent when they're creating fiction. I think the reason is that you have to have, in your own mind, a model for all of reality, so that you can place your fictitious character in a plausible world. You can afford to be much stupider in real life. When we live our lives, we can let reality determine what happens. We don't have to remember that if you go upstairs, you're no longer downstairs, or that when people get angry their faces tend to show it. If you forget, the world will remind you. But a novelist has to keep that entire world in her head.

RG: There's this tremendous amount to hold in mind that the reader isn't aware of. It's like a kinesthetic sense. You know where your body is; you don't have to look where your legs are right now. Well, a reader wanders through this reality, knowing without having to consciously think of all the things that have to be there. But if the writer violates this sense, if there's an inconsistency at all, the reader will immediately spot it.

SP: It breaks the spell.

RG: So there's a lot to carry in your mind when you're a writer. But what's so strange to me is that I've discovered things that I hadn't known and that I don't think I ever could have discovered except by being half out of my mind in fictive enchantment. Like the mattering map.

SP: Which is?

RG: Well, it's this idea that one of my fictional characters, Renee Feuer, in *The Mind-Body Problem*, came up with. One of the ways that you can understand a person is to understand what truly matters to them, what zone they occupy in the mattering map.

SP: So, in the mattering map of a typical academic, clothes count for nothing. Intelligence is the only thing that matters.

RG: If somebody tells me, "Nobody's worn that since 1983," what do I care? That's not my zone of the mattering map. Renee Feuer came up with an elaborate theory of mattering maps, and I've learned that this idea is now used in certain branches of psychology. I would say that I was the one who first formulated it, only that feels slightly dishonest. My character formulated it, or I did when I was trying to think like her, or when I was her. This state of fictive enchantment can put you in a strange relationship with your own identity, kind of hovering outside of it. That happens when you're deeply reading, too, of course, but even more so when you're writing. And that weird hovering puts you in the way of big truths. That's my impressionistic, nonscientific explanation for why writers are smarter when they are writing than when they're not.

SP: What about science fiction, or fantasy, or magical realism?

RG: There are still a hell of a lot of facts to keep straight—

SP: Yes, and I guess it's only a few circumscribed aspects of reality that are explicitly contradicted. The rest of reality, the reader assumes, works the way it has always worked. Otherwise, anything could happen, and I suspect people would get no pleasure out of fiction.

RG: The imagination is a cognitive faculty that's greatly underrated. Imagination is extremely important in science, as well. One of my favorite quotes from Einstein is something like, "When I look at myself, and my own methods of thinking, I have to draw the conclusion that my gift for fantasy has meant more to me than my talent for absorbing positive knowledge." So Einstein was a fantasist, too.

Rebecca Goldstein and Steven Pinker

Special relativity arose out of his famous thought experiment of trying to imagine himself riding a light beam. That rigorous theory originated in an act of imaginative fancy.

SP: My former colleague Roger Shepard has argued that every major finding in the history of science was first discovered through imagination, or at least could have been. Galileo could have discovered that objects fall at the same rate regardless of what they weigh by imagining two one-pound weights falling side by side, and then imagining a drop of glue joining them into a single two-pound object. Of course, that drop of glue couldn't have sped up the fall! And Roger showed that many great scientists were like Einstein. For example, Maxwell, of equation fame, began by visualizing electromagnetic fields as if they were elaborate sheets and fluids.

RG: Analytic philosophers are always making up stories, too. Sometimes the stories are very elaborate. Can you blow up a fat guy who's blocking the exit to the cave, if twenty people are inside about to die? Or, if time travel is possible, can you go back in time and kill your great-grandmother before you were born?

SP: What would happen if a lightning bolt hit a swamp, and the droplets of goo just happened to coagulate into a molecule-by-molecule replica of myself?

RG: Lots of highly imaginative stories are told. But they're not enchanting in the way that fiction is enchanting. These are stories constructed precisely to figure things out, and because of that you're never in danger of falling under a spell, stepping out of your identity. No matter how imaginative these thought-experiments are, you experience them as a form of argumentative reasoning. There's the pleasure of good thinking, at least when the thinking is good, but there's not the distinctive pleasure of fiction.

SP: In some cases, thought experiments have been incorporated directly into mediocre science fiction. *Back to the Future*, for example, was based on your thought experiment about time travel, and *The*

Matrix was Descartes's evil demon brought to life. Many episodes of *The Twilight Zone* and *Star Trek* were philosophers' thought experiments staged for the camera.

So what is the extra ingredient that a good novelist supplies? Is it some combination of the two parts of fiction we discussed—the cognitive advantages of seeing how hypothetical scenarios play out, together with the emotional pleasures of empathizing with a character to whom good things happen?

RG: I think you're right, that you can't discount either factor—although I think it's more the pleasure of feeling a life that's not your own, not necessarily vicariously participating in good fortune. Tragic art provides some of the deepest aesthetic pleasure of all. Another factor that might contribute to the deep pleasure of storytelling is that it confers significance. Stories, unlike life, have a point. We'd like to think that our lives have a point, though we often suspect otherwise. But stories are shaped around points, even if the point of the story is the pointlessness of our lives.

SP: This may connect to our original question, namely, why are we only now seeing a convergence of science and culture? Perhaps it's because science and philosophy themselves have changed in the last thirty years. They are starting to address themselves far more to the themes that ordinary thoughtful people talk about. As I recall, one of Renee Feuer's frustrations with her career in philosophy was that philosophy had abandoned all interesting questions and instead was just trying to expose fallacies of reason that came from unclear language.

RG: Philosophy just got so modest. It's laudable to try to be clear and precise, but in philosophy you have to risk a little imprecision or you're going to end up saying nothing at all.

SP: We're seeing philosophers grappling more with the substantive problems of life, like the self, morality, consciousness, after having abandoned them as meaningless.

RG: Philosophy is subservient to the sciences. It's often trying to explain the philosophical implications of the sciences, to clarify, do clean-up work. I have this terrible image right now of the guys in the circus following after the elephants, cleaning up.

SP: [*Laughs*] While philosophy was dominated by logical positivism, psychology was dominated by behaviorism, which also denied that there was any meaning to concepts like emotions, images, morality, will, and consciousness. It's only recently that the science of mind has considered those to be respectable topics again. So perhaps it's natural that fiction, which has always dealt with those problems, should only now be getting inspiration from the sciences.

RG: I think that your field in particular, and specifically your work, has had such radical philosophical implications in terms of consciousness, the self, intelligence, the moral sense. People like you have helped force philosophers back into philosophy.

SP: Even cognitive science was, for many years, devoted only to the nerdiest parts of our minds, such as recognizing shapes and stringing words into grammatical sentences. One reason evolutionary psychology has become so popular is not that it invokes evolution but that it deals with the problems that laypeople consider central to their experience but that were long banned from the psychology curriculum. Love. Sex. Family. Status. Dominance. Motherhood. Gossip. Religion. Play. Food. Beauty. Jealousy. Disgust. One might think that these would be basic topics in any science of the human mind. But don't try to find them in the psychology textbooks. Evolutionary psychology is trying to win back a place for them.

RG: The theory of the mind that you've given us is making them amenable, at long last, to rigorous explanation. I think it's a very good sign for our culture that your books sell so well. This theory is not only intellectually exciting, in the way that good science is, but it globally changes the way to regard so many aspects of human life. It's deeply transformative.

SP: Which may be why you see evolutionary psychology and cognitive science appearing in the fiction of people like Ian McEwan and David Lodge.

RG: There's a sea change, felt throughout the culture. There's a return to substantive questions, and there are artists fascinated with the thinkers pursuing substantive questions, the romance of knowledge. What we are seeing now also is more tolerance for being mystified, for recognizing that we sometimes stand, scientists too, in the presence of questions that are too big for us. We're allowing ourselves to be stunned by immensity and by our own cognitive incapacities.

SP: One of the ironies of treating the mind as a part of the natural world, as a product of evolution, is that one expects there should be problems that the mind is incapable of grasping, simply because of the way it's put together. You wouldn't expect a rat to learn a maze in which the food is placed in the prime-numbered arms—not because there's anything mystical about prime numbers, but because that's not the way a rat brain works. For the same reason, there may be limitations to the way the human brain works that make certain problems eternally paradoxical. One of the problems might be what philosophers call the "hard problem of consciousness."

RG: The "easy" problems of consciousness you guys solved. Those are the problems you take on in *How the Mind Works*, subjecting them to the powerful explanatory machinery of the modular and computational theories of the mind and Darwinism.

SP: "Easy problem" is said with a smirk, because there's nothing scientifically easy about it. But at least they are tractable. These are questions like, What's the difference between conscious and unconscious information processing, and where are they found in the brain? Why, in terms of engineering design, would a brain have evolved to segregate certain kinds of information and make them accessible to decision-making and verbal report, and to seal off other kinds in dedicated processors? We don't have answers to these questions yet, but I think we will.

Whereas the hard problem—

RG: That's the traditional problem, the mind-body problem, that we haven't solved.

SP: Yes—why consciousness in the sense of first-person subjective experience should exist at all. Why it should feel like something to be a heap of firing neurons.

RG: There's something poignant and ironic about the situation that we're in. We're inhibiting these minds, which in some sense we know very intimately. We're also learning ever more about our minds: it feels like something to have them, to be undergoing these computational processes. And why that is is still something we don't have a clue about. That's the question that isn't computational.

SP: Which, tragically, is the most indubitable thing we know. Namely, that we are conscious, as Descartes famously pointed out. So Descartes might have had the last laugh after all, even if he became something of a laughingstock among scientists, as we see in book titles such as *Descartes' Error* and *Goodbye, Descartes*.

RG: He's the general-purpose whipping boy in philosophy and in science.

SP: Because of his dualism.

RG: He isolated the hard problem of consciousness, which was good, but then he went ahead and drew an ontological conclusion that attracts derision. Seeing the obstacles to explaining consciousness scientifically, he inferred that consciousness isn't located in the body. According to him, the subject of experience isn't identical with the body and there you get your good old "ghost in the machine," as the philosopher Gilbert Ryle facetiously dubbed it.

SP: Today we would say that consciousness is undeniably a manifestation of the physiology of neural tissue. You don't need extra ghostly stuff. But we can't really explain why it should feel like something to be that tissue, perhaps because our minds just don't work in a way that would allow us to understand the explanation. Perhaps an extraterrestrial with a brain that worked in a slightly different way could understand it.

RG: It's amazing that we've managed to get as much of a grasp of the world as we have.

SP: Given that the mind is a biological gadget.

RG: Given that the mind is a gadget that evolved in an environment where we were basically trying to solve problems of hunting and gathering, evading predators . . .

SP: Prospering in social milieu . . .

RG: The fact that the mind evolved to deal with those concrete situations, and it has applied its cognitive capacities to doing string theory and figuring out how the mind works is pretty amazing. So the thought that it can't grasp everything, particularly about itself, doesn't seem surprising, does it?

SP: In fact it would seem to be necessary, if we really are evolved creatures and not angels. Only angels could understand everything about everything.

RG: And we're no angels. [*Laughs*] ∞

Rebecca Goldstein and Steven Pinker

Three

Time

Alan Lightman and Richard Colton

The author and the choreographer discuss the relationship between art and time.

Five years ago, Richard Colton was working on a dance adaptation of *Einstein's Dreams*. He learned that the best-selling novel's author, Alan Lightman, was his neighbor, and invited him to attend opening night. It was serendipitous, since Lightman was working on a novel with a dancer as its central character. Although this was not the first time Lightman had written about dance, he wanted to learn even more about life in a ballet company, and asked Colton to educate him. Lightman and Colton have continued to collaborate and are now working together on a new interpretation of *Einstein's Dreams*.

Alan Lightman received his PhD in theoretical physics from the California Institute of Technology. He is the author of more than a dozen books, including the novels *Ghost*, *Reunion*, *The Diagnosis*, and *Einstein's Dreams*, as well as the essay collections *A Sense of the Mysterious* and *The Discoveries*, which was published in November 2005. *Einstein's Dreams* was an international best seller and has inspired numerous theatrical productions worldwide. *The Diagnosis* was a finalist for the 2000 National Book Award in Fiction. Lightman has served on the faculties of Harvard and MIT.

Richard Colton is a dancer, educator, and choreographer. Colton's professional career as a performer began in New York City in the 1970s and '80s with Twyla Tharp Dance, American Ballet Theatre, Joffrey Ballet, and the White Oak Dance Project, under the direction of Mikhail Baryshnikov. In the 1990s Colton served as resident choreographer, with Amy Spencer, for American Repertory Theater, directed by Robert Brustein. Colton is a three-time recipient of the Massachusetts Cultural Council's Artists Grant in Choreography. Spencer/Colton's adaptation of the novel *Einstein's Dreams* by Alan Lightman was awarded the Sloan Foundation Science and Technology Award. Colton is the founder (1996) and director, with Spencer, of Summer Stages Dance, a nationally renowned dance workshop and performance series that takes place during July in Concord, Massachusetts, at Concord Academy and at the Institute of Contemporary Art/Boston.

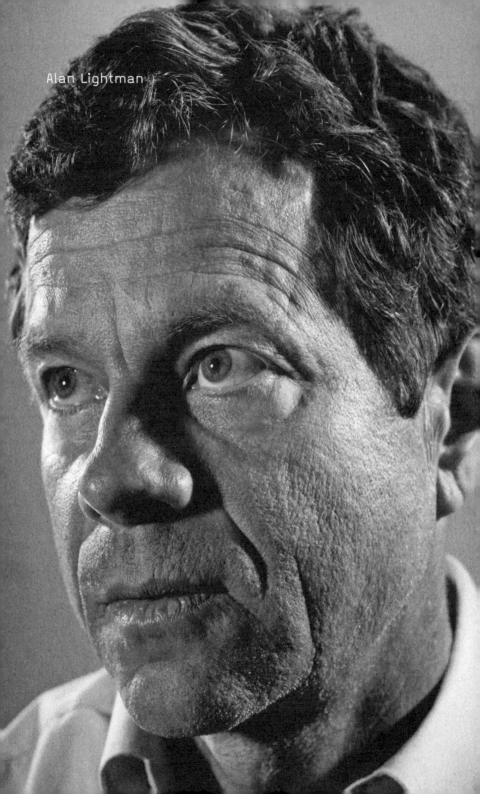
Alan Lightman

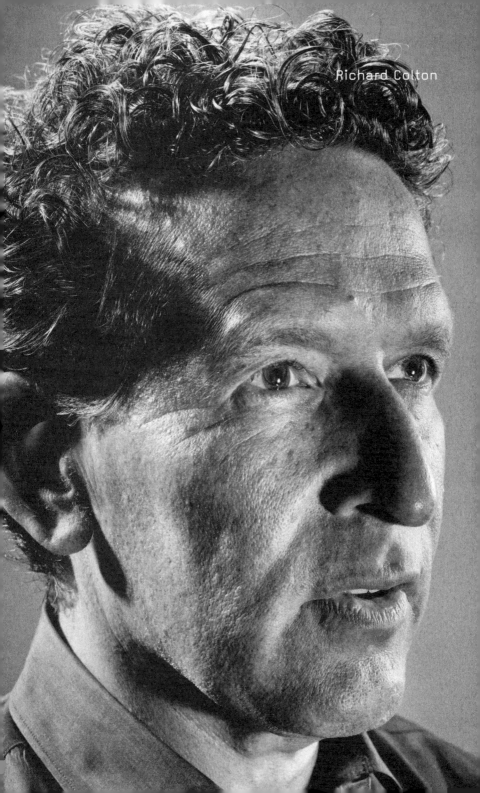
Richard Colton

Alan Lightman: I want to start by making a confession. About two years ago, I was sitting at my desk working and I suddenly found that I was able to look in at myself from the outside, an experience that rarely happens. And what I saw was that I never wasted time. From the moment that I woke up in the morning to the moment I went to bed at night, almost every minute of my day was scheduled.

If I had a few hours or longer, I could work on a writing project. If I had half an hour, I could do errands or pay bills. If I had only two or three minutes, I could answer telephone messages. I realized that I had carved up the entire day into five-minute units of efficiency, and I was appalled. I was very upset to think that I was becoming a robot—and I'm wondering, how do you use time in your life?

Richard Colton: One of the things that came to mind when you told this story is something I remember reading during a Gertrude Stein phase, which is that Stein believed that the first ingredient for creativity was boredom. You must trust that the mundane will lead to something interesting.

John Cage also taught that if you let the duration of a movement or musical phrase just keep going, it will almost always become more interesting, which is the exact opposite of carving something up into small increments. You will go through a period where the music seems boring, but if you let it keep going it can actually become quite interesting.

AL: Like Philip Glass.

RC: Yes.

AL: So you have been able to accomplish this in your life? Because I haven't in mine.

RC: No, I'm not good at this at all. But it is always something that's running parallel to this kind of madness and overscheduling and feeling that you constantly need to do things, buy more, be in step with a pace that seems to be getting faster and faster.

AL: It seems to me that one of the down sides of technology has been to facilitate this speeding up of the pace of life.

There really aren't these spaces of boredom that Stein was talking about, which I think are very important for creativity. We are just running faster and faster and faster. Not having time for reflection— not only for creativity, but also to think about who we are and what our values are and where we're going. That concerns me a lot. Those kinds of thoughts require spaces of silence and privacy.

RC: I remember when I worked with you on *Reunion*—which was a wonderful time for me—I guided you through some of the dance places in New York City. We went to this theater on St. Mark's Place and into a dressing room. For me, the dressing room is the place where you get something done. You put on your makeup, your costume. You're getting ready for something. I thought I was going to bring you there, you were going to look around, and we were going to go on. But you said, "Oh, Richard, why don't you go and have lunch. I want to be here for a while." I think it was about two hours that you sat there, and it was so interesting to me. It gave me an insight into your writer's world, your tempo of creating, your receptivity to the smallest detail.

Then, of course, reading the book, I realized this dressing room became a central place in your story. I won't go into a dressing room and look at it the same way again. I think this says something about the kind of time that is needed for the creative process to unfold.

AL: Do you feel the arts offer a respite or an escape from this sort of time-driven world that we live in?

RC: Absolutely. The arts take you outside of a mechanical sense of time, and either give you a sense of a full life's arc in a mere two hours or extend almost infinitely one brief, fleeting moment. Art's time operates at such a different rate than clock time.

AL: Yes. Whether it's listening to a concert, or watching a dance performance or a play, the arts take me out of my normal time-driven routine. I'm in another dimension. For the period that I'm watching, I'm having an experience where time is basically stopped.

In a way, it's exactly the same experience when I'm writing. I lose all sense of where I am or who I am. As the world is driven faster and faster, and we keep carving up our days into smaller and smaller units of time, I think that the experience of suspending time while watching a work of art becomes even more critical to our sanity, to our sense of who we are.

RC: A dancer like Baryshnikov, through his sheer virtuosity, can stop time. When you see him go around for the sixth revolution of pirouette, the sense of suspending time becomes a visceral experience for the spectator.

AL: But there's a paradox here that maybe you can explain to me. When you experience certain kinds of art—like movies, theater, dance, or concerts—that take place over a period of time, you totally lose your sense of time. As you said, time is stopped. It's suspended in a real psychological sense. And yet time and synchronicity are very important in dance. So there is a sort of paradox here, that you are experiencing something that takes you out of time and yet the art form itself has to be choreographed. It depends on time. When you are choreographing a piece, how important is the timing to you?

RC: Well, this is really interesting to think about. When I get into the studio with dancers, is it space I'm dealing with first, or time? I've come to the conclusion that time is primary. Duration, how long something lasts; tempo, the speed at which something takes place; kinesthetic response, which is your reaction to some outside movement or sound; repetition, either of someone else's movement of something you've done—are all time-based.

I think it's in Wagner's *Parsifal*, when Wagner describes the setting of the Holy Grail, that he says, "Time becomes space; time is primary."

Balanchine, who is probably the greatest of twentieth-century choreographers, always needed music first. He said his only requirement to begin his work was having good music. He couldn't work simply in space. He needed the time element. Now what you're describing, I think, is the tension that's in all of the performing arts—the contradiction of pulling away from time while also depending on it for structure.

All great dance pieces—Twyla Tharp's *Fugue*, Tricia Brown's *Glacial Decoy*, Merce Cunningham's work—do this sort of weird thing with time. There's this natural forward momentum, speed, but at the same time it suddenly arrests and turns away from itself. And it's that tension—maybe as a metaphor you can think of blending future, past, and present—that kind of playing with different dimensions of time, that dance is wonderful with. Dance can visualize beautifully how time moves the same for all human beings, but each human being moves through time differently.

AL: You mentioned that you feel that time precedes space in thinking about what you're doing in dance—and yet, at the same time, it is also used to map out where you are in spatial relationship to other dancers.

Time was also used to indicate space in the eighteenth century. One of the main problems for navigation was knowing where you were in terms of longitude. This was a practical problem. If you didn't solve it, you would have a shipwreck. So if you were trading, not only money depended on it, but lives.

It's easy to determine latitude. You can do that astronomically by just one sighting. But to determine longitude, you have to have two clocks that are synchronized with each other. So it became a major scientific problem: How do you create clocks that have enough accuracy that they can maintain synchronization while going over the bumps of the ocean? That's a very explicit example of time being used to measure space.

I think time and space have been tangled up since then. When Einstein came along, he just found a more profound way in which time and space were tangled together—but in practical terms, they had been tangled for quite a while.

RC: John Cage, Merce Cunningham, and Robert Rauschenberg created a major revolution in dance by pretty much throwing out everything that had held the stage together for centuries: narrative, a classic stage perspective, musical synchronicity. The only thing they didn't throw out was time. They decided that each of them needed to know how long each section was. They used a stopwatch to do this, to create their American type of dance, they did keep time.

AL: I think our sense of time, our sort of visceral understanding of time, has been one of the most permanent aspects of our experience of the world. So much else has changed—but there seems to be something so primitive about our experience of time and what it means. I think it may even start in the womb. I mean, one thing that most philosophers agreed on since earliest recorded history was the absolute nature of time.

RC: And then that got overthrown, that sense of the absolute.

AL: But even Isaac Newton, whose own equations of motion had a relativity principle embedded in them, believed in an absolute. He believed in an absolute time and an absolute space that were created by the body of God. Like most everybody, including intellectuals, he was religious. But that sort of absolute nature of time was something so deeply ingrained that it really was the one unquestioned thing about science until Einstein.

On the other hand, when one talks about the psychological experience of time—and I think it's very interesting to contrast the psychological experience of time with the physical nature of time—we know that time is experienced in a lot of different ways. This may be part of the resolution of this paradox that I was talking about, between the way time seems suspended when you watch an artistic

Alan Lightman and Richard Colton

work and the actual mechanization of the choreography that involves time. One is more of a physical time, and the other is more of a psychological time.

I wish that I had been around when the first clock was invented, to see how people must have reacted, realizing that there was a time that flowed outside their bodies, a time that could be marked off and measured. It must have been a very strange experience to be confronted with something that you felt before, and were now suddenly to be presented with a sort of a quantitative measure of this elusive thing.

RC: To me, the kinds of contradictions you talk about are exactly what the arts deal with so powerfully: the difference between human time, mechanical time, time that is still, time you're trying to find and freeze, different kinds of time occurring simultaneously.

AL: I want to ask you another question about choreography, because using the human body to create art is something that amazes me. When you choreograph something, do you use a watch? Do you ever explicitly look at how long it's going to take for one of your dancers to get from A to B if they're doing a step?

RC: There are choreographers who do that. Merce Cunningham always seems to have a stopwatch with him. I don't. I usually have specific music that I've inhabited long before I go into the dance studio.

AL: And is that music playing in your head?

RC: Yeah, I've usually been working with it or listening to it for so long that the lengths of time and duration are in my head.

AL: I know there must be something corresponding to perfect pitch but relating to time, where a person has sort of an absolute sense in their head of how long a second is. Do you have perfect pitch in time?

RC: No, no—as a matter of fact, it may have been hard to dance with me in my performance days. I was the kind of dancer that didn't really keep a steady beat.

When I'm watching a dancer now, if the dancer takes me some-place, I want to extend that moment choreographically. I'm more interested in finding the visceral energy, the kind of emotion that happens outside of coordinating moments.

AL: But when you're asked to create a piece, aren't you given a time length: "This has to be thirty minutes, this has to be forty-five min-utes."

RC: Yes. I'll often find a piece of music that's, say, twenty minutes and know I have the container.

AL: So in that case, the music would serve as your clock.

RC: That's right—and how I work with that becomes the energy and life of the piece.

AL: One of the exciting things to me in looking at the history of art in the twentieth century is seeing how it moves from the visual to the temporal. Even for radical artists like Cézanne and Picasso, their strength is still in the visual. And then you move toward Barnett Newman or Richard Serra.

RC: Yes. You start to be in this wonderful temporal world where it's about the spectator and the work of art as two moving objects, and then a third spectator watching those two interact. It's the closest I've come to understanding relativity in any kind of visceral way.

AL: I think that Einstein certainly has influenced modern artists. There are a couple of sculptors, like Richard Serra, who do these big installations—or these earthworks, as they call them—where you have to actually walk through the exhibit, and it might take you ten or fifteen minutes. Unlike looking at a painting, where you might stand a long time but it's sort of an instantaneous interaction, to experience some of these earthworks you have to walk through a quarter-mile exhibit. Some of these sculptors have explicitly cred-ited relativity with being an influence.

I know a number of writers have; Nabokov wrote a short novel called *Ada* in which his main character, Van Veen, has this long monologue and talks about relativity theory and his understanding of time. He's very upset by relativity because it seems to pin down time too much, even though it's relative. It sort of quantifies time for him more than he would like.

I don't know whether Virginia Woolf read Einstein, but in her novel *Mrs. Dalloway*, which is one of my favorite books, there's this wonderful counterpoint between mechanical time and body time. That novel takes place entirely in a single, twenty-four-hour period. I'd never before experienced in literature the close examination of what it means to be aware of the moment as in that novel. You really feel what it means to be alive and conscious, which is a very subjective, timeless experience. And yet in her novel, every couple of hours during this day we hear Big Ben, the clock, striking, which marks off mechanical time. The regular songs of this clock are like a scaffold of human-made time occurring outside of our bodies, at the same time as these flights of consciousness and timeless interior life.

We were speaking about sculptors who use time in the way that they did space in an earlier era, but the novel has always been an art form that is an experience through time. A novel is not a haiku. It takes several hours or longer to read, and during that experience you are in a timeless state.

RC: Probably the most beautiful book I've ever read on light was also by Virginia Woolf, *The Waves*. Each chapter begins with a dazzling description of sky and its light. I thought it was astounding. So she's written the greatest books on light and time.

AL: She's got it covered.

RC: Since we're talking about light and time, tell me about that wonderful image of Einstein on the light beam.

AL: It had been known for a long time, since about 1870, that light is an oscillating wave of energy that travels through space. In his autobiography, Einstein described first thinking about time. He said that when he was sixteen years old, he imagined trying to run alongside a light beam—in a sense, making it stop—to see what such a frozen oscillation would look like.

He realized that no one had ever seen such a thing, and that led him to the idea that maybe it was impossible to run alongside a light beam—that no matter how fast you were running, it would always be faster. And that, I think, was part of the thinking that led him to his main postulate of relativity theory.

A light beam always passes any observer at the same speed, no matter how fast that observer is running. That violates all common sense, because common sense tells us if you run toward an oncoming car, it's going to come at you faster than if you run away from it. But Einstein says if that oncoming car is a light beam, it will come at you at the same speed whether you are running toward it or away from it, which is a pretty outrageous idea. Yet from that postulate came all of the theory of relativity.

RC: Do you think it's the outrageousness of his idea that has influenced artists?

AL: I think artists like to have their world thrown upside down. That's part of what art is about, in my opinion. Artists like to look at things from totally new perspectives. That's why artists have always enjoyed staying in touch with science. There's a long history of salons and groups of both artists and scientists, going back to the Lunar Society in England in the 1700s, and earlier I'm sure. Artists very much like to get new ideas from science, because they shake up their worldview, which is what they're trying to do with their art.

Einstein's theory of relativity—and Heisenberg and Schrödinger's ideas about quantum theory, which are also very deeply disturbing in a visceral sense—are rich material for artists. It may be an impossible question to address, but do you feel that Einstein influenced any choreographers?

RC: I know many choreographers who have read Leonard Schlain's *Art and Physics*. In the nineties, it was passed around by quite a few:

Paul Taylor, Dan Wagoner, and others. So, yeah, the interest was there.

Science is at this particularly fascinating place. It may be more mysterious than anything we're doing in art at this moment in history. That's a tough thing for artists to feel—but nonetheless, I think artists continue having insights into the human condition that are invaluable.

AL: Absolutely. I think that art influences science, too. Especially in the twentieth century, science has gotten so far beyond human sensory perception that we're really talking about very abstract things—subatomic particles and wavelengths of light, which we can't see; the Big Bang, which we can't experience; and distant galaxies, which we really can't touch. And yet when we talk about physical phenomena, even in these inaccessible domains, we have to use language. Because that's all we have. I think one of the things that art helps provide scientists with is the language—and the metaphors and the images—to describe what scientists are so desperately trying to understand. Our instruments tell us that these totally unimaginable phenomena are happening, and yet we have no intuitive understanding of them. So we grope for language and pictures, and I think art provides some of these for us.

RC: I met you as a writer, so I think of you as a writer first and foremost, but I know you have a strong science background. Is the pull of the scientist there when you're at the writer's table working on a novel? Would you define your novel-writing and science as two separate worlds, or are you holding them beautifully in balance? What is that relationship? I'm interested.

AL: Certainly, I'm not holding them beautifully in balance. I'm struggling with them. But I've been interested in both the sciences and the arts from a very young age—and I've noticed that there are a lot of similarities, and some differences, in the way that scientists and artists view the world, and the way that they work.

One of the differences is that scientists are always working on problems that have definite answers. Even though science is constantly revising itself—you get new experimental data and you change your theory—at any given moment, a scientist is working on a problem that has a definite answer. These are so-called well-posed problems.

When I was a graduate student in physics, my thesis advisor told us not to waste time on problems that didn't have definite answers. It might take us three years to find the answer, it might take us ten years, but we had to know that there was a clear answer.

I think for artists and for most other people, there are a lot of interesting questions that don't have definite answers—like what is the nature of God? Or what is the nature of love? Or would we be happier if we lived to be a thousand years old? These are all terribly fascinating, important questions, but they're not scientific questions.

There's this wonderful quote from the poet Rainer Maria Rilke, when another young poet wrote to him and said, "Do I have what it takes to be a poet?" And Rilke wrote back, "You should learn to love the questions themselves like locked rooms, or like books written in a very foreign tongue." I think that a lot of art is about the questions themselves.

The question is more important than the answer. So I think artists are much better at living with uncertainty.

Ambiguity is an essential part of art. Again, I'm speaking part as an artist and part outside of art. I know when I'm writing a novel, I never want to understand the character fully. In fact, if I've written a good character, I can never understand the character fully. Once you understand the character of a novel fully, the novel is dead for you. So I think that one of the big differences in the sciences and the arts is that in the sciences, you're dealing with questions with answers: You're dealing more with certainty. In the arts you're dealing more with questions that don't have answers. I have sort of confronted this myself, getting back to your original question of . . .

RC: What is the struggle?

AL: For me, one of the problems that I sometimes have as a novelist is that I want to control my characters too much. I want to plot them too much. I want to know how the thing is going to turn out. It's a big danger in writing fiction to overplot your characters. You need to give your characters enough freedom to breathe and live, and that means that they may take actions that you aren't anticipating. You have to go with your characters, you have to listen to them, and I sometimes find this difficult. It sort of violates all of my instincts as a scientist. As a scientist, I want to know where things are going, to have more control, to see everything as some organized picture, where there's a problem and a solution. But that doesn't work very well in writing a novel, so I feel this conflict with myself.

A very microscopic version of the conflict is in comparing expository writing to creative writing. In expository writing, you always start each paragraph with a topic sentence that sort of names the idea of the paragraph, that organizes the paragraph for the reader. In fiction writing, a topic sentence is fatal. You don't want to tell the reader how to organize her thoughts when she's about to read a paragraph. You want her just to be carried away, to be part of the sensual experience, to participate in the creative experience.

RC: That's interesting. It reminds me of theater directors saying to actors, "Do something you think the character wouldn't do." Very often, that's the beginning of a kind of reality that breathes.

I love being at the early stages in a theater rehearsal, when actors are developing the physical life of their character, as opposed to what they're simply saying. They often create a very different physical life from what the character's language may imply, creating a sense of instability, contradiction, ambiguity.

In a way, it seems like what you're describing is the novelist's need to do all of this at once. You need to be the writer, director, and actor. It's a complicated, layered process that you're talking about.

AL: Yes.

RC: I'm glad you are mentioning all my favorite writers. No one talks as clearly and passionately about the role of art in one's life as Rilke. He wrote that about living badly because we come into the present unfinished and how art can change that.

AL: Yes—but I would argue that, at least speaking personally, I don't ever want to be finished. I would like to always be in the process of finishing myself but never complete the process.

RC: But I think Rilke was talking about the ability to be receptive to all that's around us, less distracted. It doesn't mean giving up the process of becoming.

AL: Right. It's like the concept of the beginner's mind in Buddhism. The ideal is always to be a beginner at everything, because when you are a beginner, you're receiving the most. You're not coming in with preconceived notions. I think this is also why a lot of scientists do their best work at a very young age: because they don't have a lot of tradition and dogma about what the known laws are, and what can or cannot be questioned.

RC: Like Einstein.

AL: Yeah, he was twenty-six. James Maxwell was totally finished unifying the equations of electricity and magnetism in his mid-thirties. Werner Heisenberg was twenty-three when he came up with quantum theory. All of these people had beginner's minds, in a sense. They were coming at something fresh, which I think resonates with Rilke's point that you want the present to be unfinished. You want to be able to receive and be open.

RC: A performer who is very nervous doesn't have as many options, because they're not open to all the possibilities. It's hard for me to believe that Einstein could have come up with his theory of special relativity at age forty-five or fifty.

•　　　•　　　•

RC: But I think of Einstein as a wonderful outsider. He wasn't in the circles, going to the conferences. He was thinking outside of the world of physics.

AL: That's right, but he also understood the known physics of his time.

RC: And that's often true with artists, too. There's a quality of being outside something, so you're not pulled by a kind of herd mentality.

AL: I think we might be at an end here.

RC: They say that after three minutes you start lying anyway.

AL: I think three minutes is generous. ∞

Four

On Design
Drew Endy and Stefan Sagmeister

The graphic designer and the Stanford synthetic biologist discuss
beauty, elegance, and the requirements of good design.

Stefan Sagmeister and Drew Endy were both invited to Los Angeles to participate in Image and Meaning 2, a conference about visualizing science. Endy, a scientist who designs, and Sagmeister, a designer interested in science, were well-suited to broach the use of imagery in science, and what art and design can bring to the scientific table. They caught up again in New York City and took the conversation further, to discuss the act of creation, the usefulness of beauty, and all things design.

Drew Endy is a biological engineer at Stanford University who develops foundational technologies that help to program DNA in order to design and build many useful living systems. In 2004, Endy cofounded the MIT Synthetic Biology Working Group and the Registry of Standard Biological Parts, as well as organized the first international conference on synthetic biology. Endy also cofounded the BioBricks Foundation, a not-for-profit organization working to promote the open and constructive development of biological technology.

Graphic designer Stefan Sagmeister arrived in New York as a Fulbright Scholar in 1987. After leaving to work for the Leo Burnett agency in Hong Kong, Sagmeister returned to New York City and joined the celebrated M&Co studio under Tibor Kalman. In 1993, he established Sagmeister, Inc., and went on to produce his best-known work, including album covers for Lou Reed, the Talking Heads, and the Rolling Stones, as well as the now-iconic posters for the American Institute of Graphic Arts. He is a four-time Grammy nominee and the winner of a 2005 National Design Award, as well as the designer of *Sagmeister: Made You Look* and *Things I Have Learned in My Life So Far*.

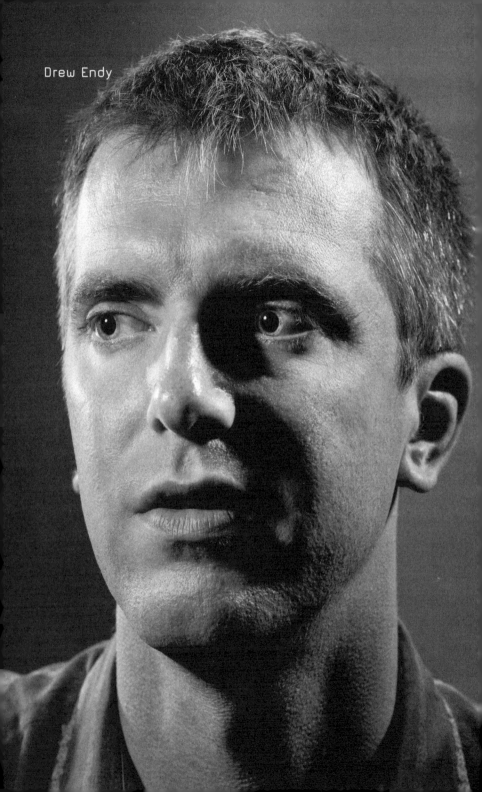

Drew Endy

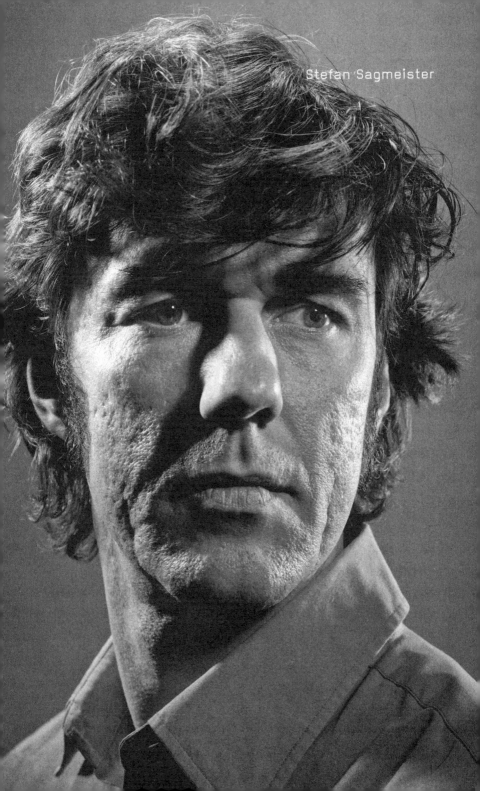

Stefan Sagmeister

Drew Endy: I'm an engineer working in biology, which means that my interests are to design and build biological systems—not like most folks working in biology, who work to understand nature's designs, but rather to construct new things that don't exist. And because that problem is so frustrating and hard, I spend most of my time thinking about, and trying to invent, foundational technologies that would make it easier to do that.

Stefan Sagmeister: It seems to me, from what I understand you do, that you are very much a composer while the guys who do genetic engineering are more like DJs. Is that correct, considering that you build elements from scratch while the genetic engineers take different segments and sort of mix them up?

DE: They are making mash-ups, yes. Though I'm not so much a composer as the person who is trying to figure out what the scale looks like and what the notes are, so that somebody, later on, could use this information to do composition. Because right now we have no idea how to do that.

SS: So in that sense, the DJs are quite a bit ahead of you?

DE: Way ahead. I mean, *way,* way ahead, and will be so for quite a while, which is fine. The real issue we're facing right now is that we have no idea how to design and build biological systems; we're clueless. We've got 3.6 billion years, approximately, of evolution, and it's beautiful and amazing. But if you gave me a cell and I looked at it, I wouldn't understand it—not at the molecular level; I wouldn't know how to use the DNA to design one myself. And so I wouldn't be able to say whether it was good or bad or ugly in its design. It's beautiful because it's this living thing, and there's something about that which is magical but—

SS: So, a cancerous cell is not somehow less elegant looking than a healthy cell?

DE: It depends on your perspective—I mean, superficially, explicitly, from the value of somebody you care about or love, it's ugly and horrible. But as a replicating system within an environment, which it may be parasitizing or destroying, it's amazing. So there's the question of when you evaluate a design, what makes it a good or bad design? What is the system that creates a value around it?

SS: I think it's, as you say, completely perspective- and context-dependent. For example, in graphics, the swastika was a wonderful symbol used in many cultures around the world for hundreds of years, until Hitler came along and changed the context.

DE: Right. So there are these questions. I have lots of questions for you, because I'm in a situation where I feel like I don't know how to design, but I'm trying to invent means of designing; like maybe I'm in the business of begging, borrowing, and stealing ideas from other human experience with design, because we're trying to learn how to do it in this particular substrate.

SS: Ask away.

DE: Well, when we design the inner working of a living cell, at the molecular level, our only audience—the only viewer—is the cell itself. So our designs are evaluated only at the level of function, and not by how they "look." So I'm interested in—maybe jealous—when I see that you get to convince somebody to pay attention to design, to select for it, by mapping a message onto a carrier that attracts attention to the design and its ultimate message. I don't know if there's a word you use for them—where you're mapping onto a pay-attention-to-me, or check-this-out, and then now that I've got your attention, I'm going to give you the real message. It's a Trojan message, and I'm wondering if, and how, we do that inside living cells.

SS: Yes, of course, in graphics we create that all the time: I'll show you a chicken with its head cut off to entice you to read about a conference program. So yes, we use check-this-out, attention-getting devices to get the message across. As for whether this happens on a molecular level—in general I am actually not so fond of the Trojan message; I prefer straightforward communication.

DE: But I would think that straightforward communication is so much more challenging.

SS: I would say more honest, not more challenging. And I find honesty almost automatically interesting.

DE: So when you do use the attention-getting devices, is there a formal modularization to the design process where you go, here's a column of things I could use to catch people's attention, and then, here are the messages I want to communicate, and I'm going to optimize the mapping between pay-attention-to-me and think-about-this?

SS: Actually, I never put it down into consecutive steps and my guess would be that if I tried, it would probably produce a pretty good design, but not a great one, because, I think in the end, a lot of these decisions are actually made with the gut. I was thinking quite a bit about this world of design that can touch somebody's heart. And of course, there are various ways designs have touched my heart, and I can pinpoint different things that they have in common—they might have taken a lot of guts to do and those guts are still embedded in the final visual; or they might possess revisitability, something that made me look at them again and again. Or it might just be incredibly beautiful. It might have had an element of surprise to it. But of course, in design, "surprise" is, many times, simply just shock.

DE: Something used for shock value.

SS: Yeah, so you could check every design you do to see if it incorporates these things. My guess would be that you would wind up with something that is closely related to the results you would expect to get from one of the self-help books at airports. You might not wind up completely off the mark, but the likelihood that you'll come up with something true is slim.

DE: You mentioned beauty, how important is that? Or aesthetics, or elegance—how important are those to you? Speaking for myself, I couldn't design something that I didn't want to design. And part of wanting to design something is the hope that it's an elegant, beautiful thing—having the motivation to do it. You could motivate somebody by withholding food—I don't know if people have done those experiments with designers—and see whether or not good work comes out. But if there is self-motivation involved, then part of that tends to lead toward celebrating something beautiful.

SS: Actually, I really changed my mind on the question of beauty and elegance because I used to have a boss, Tibor Kalman, whose little line was "I have nothing against beauty; I just don't find it very interesting." And we ourselves, in the very beginning of the studio, had this line up on the wall that said, "Style = fart," meaning any stylistic questions are hot air and have no meaning whatsoever. And experience proved me wrong. I found that if a message is to be communicated, even from a purely functional level, style and beauty will play an important role in the communication of that message. And in that sense beauty works; it has a function.

DE: For me, if it functions at all, it's beautiful.

DE: I've heard you refer to finishing designs as releasing them into the wild, right?

SS: Yes.

DE: You do something and out it goes, and you've tried to consciously make sure that the design is something that is good, constructive, adds something to what's already out there.

SS: We try to, yes.

DE: So have you ever been surprised by the way something plays out? Or can you usually forecast, with some accuracy, what will happen when you complete a design and release it? Because in biology, say for something like genetic modifications of plants, there are often predictions, and then you put it out in the field and something else happens. That can be good, but it isn't always good.

SS: Yes, sometimes when you put something out there reactions occur that you don't expect; "the field," as you call it, reacts to the design and can even change it in a way. When we designed a page for Lou Reed, for his CD cover, we used a picture of him with scribbled handwriting all over his face—this idea came out of a desire to illustrate the personal nature of the lyrics. Lou liked it a lot so we designed the poster for the CD in a similar fashion. The poster went on to be published in basically every design and advertising magazine around the world, and so it makes me think of what you're saying about genetic modifications, because this image had an effect that we never expected. In a way, it infected design; it spawned this incredible scribbled-handwriting infection, I would call it, where even IBM ads started to have scribbles all over them.

But you know, the climate at the time—the very formal new Modernism, with lots of slick and cold graphics—was, I think, ready to be infected.

DE: Right, right. For an infection to spread to a lot of similar hosts—for it to propagate—you need a monoculture. It sounds like you infected a monoculture. And that reminds me—I was in China earlier this year and I saw some pieces at a gallery where there were characters scribbled over people's faces—I wonder if they were descendants of Lou Reed?

SS: I would not be surprised; I saw iterations of it from Paris to Caracas. But I'm interested in this idea of prediction for someone like you. I would assume that this question is a very big one if you're involved in synthetic biology, yes?

DE: Yeah, it's embedded in ideas about how to navigate a path. Like, when you start to enable design and it's going to happen one way or another, and so much as anybody can help bias the direction of a gigantic snowball that starts rolling, you'd like it to be biased in a direction that's constructive. But it's not clear how to actually do that. It's sort of like, I couldn't now tell you everything I want to design in the future—we'll have to see what the future is like. And so I also can't know everything that I might need to do, or want to do, to make sure that I'm being responsible—which is such a huge part of prediction on my end of things.

It sort of comes down to the idea that there's something about human existence that celebrates construction, that celebrates more life, more experience, more beauty. And if you can't rely on that, ultimately, we're screwed. As we start to engineer biology systematically, we have to make sure that we continue to celebrate what it is to be a living, constructive society. There are lots of technical details and caveats, like, "Don't put a gene that codes for a toxin into a bacteria that lives in the human gut." But all seem to be motivated by this much more fundamental thing, which is part of human nature: to make stuff. It's a celebration of living. And, I've figured out, that's what drives anything for me, which has to do with responsibility.

SS: Well, that is beautifully put; I have nothing to add to that.

DE: Do you ever look at things in nature, biological stuff, and think that it's good or bad or beautiful? Not so much at the level of existence, but at the level of design. So that the superficial example would be, oh, hey, check out that giraffe, it's got the long neck. Cool.

SS: Yes, I'm a huge jellyfish fan.

DE: Have you ever been to the Monterey Bay Aquarium?

SS: I have been to Monterey Bay, but I think a highlight of my life was a special exhibit they had on jellyfish in Baltimore, which was just gorgeously put together, and they had the widest variety of species.

DE: So living things from nature, are they always beautiful? And if you had to, could you start to evaluate designs in nature, and could you say, this is good, this is bad?

Because the problem we have in biology is that we can work on the hard problem of figuring out how it actually works, but we don't have a value system, which lets us say, wow, what an elegant design.

It could be an incredibly impressive molecular motor or something like that, but how does one choose to evaluate a design? It could be aesthetics. I guess some people would say function; some people would say the efficiency of the encoding of the information specifying the design. I mean, you could invent characteristics.

SS: I think the evaluation system for a piece of graphic design and the evaluation system for a piece of biology are not really that different. I think you could probably use the same systems. I can definitely tell you how I would evaluate a piece of graphic design.

DE: And also a jellyfish in the aquarium—?

SS: It would be subjective, but I would treat it very much like a piece of graphic design. I would want to ask what does it do and how well does it do it? Have I seen something like this before, or is this completely novel? Is its form, and maybe even its style, coming out of its function, and do they work together as a beautiful whole? Is it a necessary thing? Is it good for other things—

DE: Is it food, do you mean? Would that be necessary?

SS: Well, is it necessary for other organisms to survive? Or in the case of graphics, is it something that has a right to be in the world? Which I guess you could put forward for both systems.

DE: They're surprising and they're beautiful, and I don't know how they work. That's something which is part of the novelty, I guess.

SS: Yeah, the mystery, which is something I think many artists get quite a bit of mileage out of; the fact that they work in techniques where the general viewer has no idea how such a thing would be done. And that contributes to the overall magic of the object.

DE: And that sort of value is in part derived from a decoupling, where the viewer, the person who is experiencing the object, can't imagine, or, as you are saying, doesn't know how you would do that, right, or doesn't believe that they themselves could do that. And so that is, to an extent, how we're currently interacting with the living world, in that we've got these beautiful artifacts and we can observe

Drew Endy and Stefan Sagmeister

them, they're evolving, we inherit them from the living world and celebrate them and they're mysterious because we don't know how they work. I wouldn't know how to write down a piece of DNA—A, C, T, and G—that programs some cell to grow up into a jellyfish. As a biological designer, until I can actually design something, I don't understand it.

And now we're at this weird point where, even though we don't understand everything, in fact most things, about the natural living world, we're beginning—via technology standardization and powerful paintbrushes, like DNA synthesis—to create, to construct. And so we're implementing a collapse, basically, of the decoupling; we're starting to participate and systematically design biological systems. And thinking about doing that is incredibly scary from some perspectives, mine included, because we don't know necessarily what we're doing—we're just going out and trying it.

SS: Yes, it's interesting that, on the surface, you would think you could get an understanding of something just by looking at it, or experiencing it. About a year ago, I read an interview in the *New York Times* with a young boutique hotelier in New York who answered the questions about his qualifications for opening and running such a hotel by saying that he has stayed in many of the finest hotels around the world. I have never heard of him or the hotel since. So, yes, I agree, you really need to have designed something to fully understand it—I really didn't understand the structure of a music video until I directed one. But if you're involved in designing living elements, it seems to me to be a matter of much more profundity.

DE: Why? Because they're living? I run into a curious problem, a reaction I get from fellow researchers, biologists, physicists when talking about designing biological systems. They say, "Biology is not like that." And what that means usually is there is something magical, there is something enchanted, about life—and certainly there is from an experiential perspective—which doesn't apply to biology. But I think we need to disenchant the molecular details of biology without destroying its magic. So that we begin to navigate a path where the idea of design is accessible, but a sense of awe is maintained. So that when we get to the point where designing biology is no different from anything else, from any other form of design, we're in a world where the vast majority of the designers are doing good.

SS: I think if I were given the opportunity to do biological designs, I would be scared into an absolute inability to do anything by the incredible unpredictability of the outcome—maybe now I'm getting a sense of the responsibility you talked about! But, yes, everything you're doing is so incredibly brand new and mysterious; you're really laying the foundations. The systems in which I work are very much explored. The field is at least one hundred years old, so we can stand on the shoulders of the designers before us. We're basically all fishing in the same river, and we often come up with very similar fish. And it is a challenge for designers, because I think the brain has an exceptional capacity for repetition, for doing things over and over again, which of course is good; we could not survive otherwise—we'd need to think every single time we drank a glass of water about how to stretch out the arm to grab the glass. And when I am trying to come up with something new, I have to think of ways to break that pattern, of techniques to throw my brain off its beaten path. For example, when generating an idea for a logo, instead of thinking about it traditionally—considering the client, their history, their competitors and their target audience—I find it often more fruitful to start thinking about that logo from something completely unrelated I see lying about, say a computer cable. By the end, my logo will have nothing to do with a computer cable but since I started from a different point, the chances that I will arrive at a new solution increase dramatically.

DE: Right, and I get into the same ruts, except the problem for us is not the beaten path as much as no path. But actually, I'm really surprised to hear what you are saying, because in biology we have the scientific community and within that, the biological research community, and there is tremendous fishing in the same fishing holes. You're competing to discover something and if you're discovering something that already exists, somebody's going to be first. Whereas with design, I always thought, it's a creative, constructive process and it's therefore open-ended, and so it isn't necessarily most important to get there first so much as it is to get there beautifully. And it's true that, because my design community is quite small, you could go anywhere and throw in your fishing line and you'll be the only person fishing right now. But if what you're saying holds true, as my community gets bigger, that's going to be really disappointing. I've sort of naively been hoping that because it's a constructive, forward design process, it will produce diversity and people will be able to find their own fish, or decide they'd rather catch butterflies.

SS: No, there is one river, and we're all on it. Including, surprisingly, those doing the client-free experimental stuff. Of course, you could imagine that an identity for an international corporation is designed similarly in Caracas as it is in Oslo, but even looking at experimental books, say, one from Scandinavia versus one from South America, they all look the same now.

DE: So why do you think that is? Do you think it is perhaps because we are all people? Do you think it is because the environment, which selects or evaluates designs, is too similar? Or do you think it's the technologies that people use, the standards, the aesthetic systems, whatever, that celebrate some sort of conformity?

SS: I think it's all of the three. When my dad grew up in Austria, he didn't really know what the village three towns over really did; because of the mountains, he hardly ever visited it. But today I could trace so much of what is happening in experimental designs back to the same couple of people. Perhaps we are all too connected and so we've become, as you say, a monoculture.

DE: So if we started to design people, I don't know if it would be good or bad that everybody is fishing from the same hole; it just sort of defines how things are. Do you think that is good or bad? Do you like the fact that you are sometimes worried that somebody else might do the same thing first? Is that a feature in the sociology of graphic design or is that a bug? Because if we could start designing people, then maybe we could change the sociology.

SS: It's a bug. I think more exciting things could be done, more valuable things, if we would work in different directions—maybe you can let me know when you start designing people? ∞

Drew Endy and Stefan Sagmeister

Five

Objectivity and the Image

Joan Fontcuberta and Ariel Ruiz i Altaba

The artist and the biologist/artist discuss photography, objectivity, and the poetics of truth.

Joan Fontcuberta and Ariel Ruiz i Altaba met in 2001 through their mutual publisher. Ruiz i Altaba had been a longtime admirer of Fontcuberta's work and asked if he would write the text for Ruiz i Altaba's photographic project *Embryonic Landscapes*. A year later, Fontcuberta asked Ruiz i Altaba to curate an issue of *Photovision* on the theme of "science and art," which was titled *Genes: Identity and Image*. Here, they continue their dialogue and debate the authority and poetics of science.

Joan Fontcuberta is an artist, theorist, and professor. He has written and edited many books on photography and has taught at Harvard and the Art Institute of Chicago. In 1980, he cofounded and became editor of the magazine *Photovision*. Fontcuberta's photographs and digital images explore the boundaries of knowledge, memory, and science. His work has been collected by MoMA in New York, the National Museum of Modern Art in Paris, and Barcelona's Museum of Contemporary Art, among others. He is a professor of audiovisual communication at Pompeu Fabra University in Barcelona.

Ariel Ruiz i Altaba is an artist and a scientist. He earned his PhD in molecular biology from Harvard and is currently the Louis-Jeanet Professor of Stem Cell Research at the University of Geneva. His scientific research focuses on cancer, stem cells, and brain development. Ruiz i Altaba's photographs (www.ruizialtaba.com) explore the intersection of art and science, as well as issues of identity; they have been exhibited and collected by museums and galleries around the world.

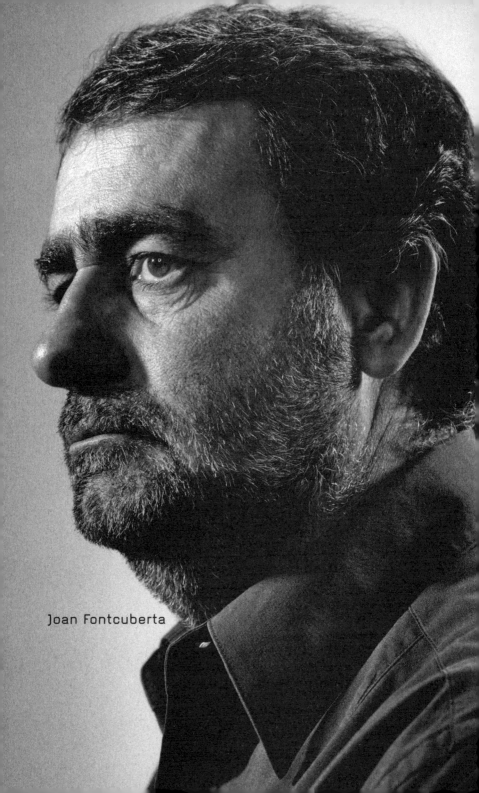
Joan Fontcuberta

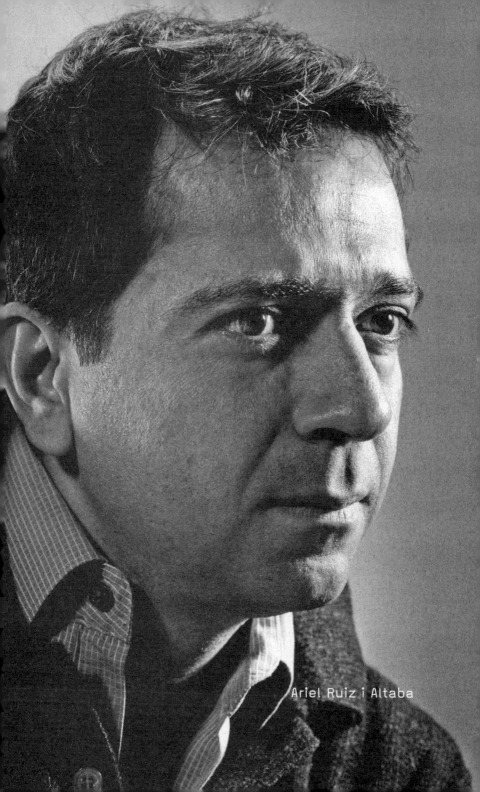
Ariel Ruiz i Altaba

Joan Fontcuberta: We had a chance meeting in Paris because of Paris Photo, where both of us were exhibiting. You were presenting some samples of your recent work *Minimal Landscapes* and I was presenting some pictures from my *Orogenesis* series. And I thought the fact that both of us were dealing with landscape was interesting.

For instance, when we talk about landscape, are we dealing with the same concept? I've known your work for a while and I've seen your evolution, and that your idea of landscape comes impregnated with concepts from biology, or that it comes from visual and graphic elements which rise from the biological iconography, because that's your starting point. Whereas I come from semiotics, from mass media studies, and other related fields, so I'm more interested in the representational issues.

Ariel Ruiz i Altaba: Yes, in both cases I think we try to stay far from what may be perceived as consensus reality—with you, because you are creating an alternate one; and with me, because I'm sort of perverting the ideas of documentation in science.

In my case the evolution you talk about has certainly come from observing the biological landscape on a greater breadth of scale, with different tools. But that hasn't always been related to questions of what a landscape is.

One thing that I thought was valuable from my previous *Embryonic Landscapes* work, which was not documentation but rather aesthetic exploration, is the multiple levels of information that one can find in the work, depending on the viewer—and therefore the participation of the viewer in creating a particular landscape.

JF: Yes, and I think another important aspect of the viewers' reading is context. For instance, pictures in the context of a scientific publication would have a specific meaning, but the same pictures published in an art book or displayed in an art gallery would have a completely different reading.

And when we deal with scientific iconography we're applying the surrealistic strategy of displacement—the fact that different elements out of their context will achieve a completely different meaning.

ARA: Absolutely. Certainly this is something that is very clear in your works *Fauna* and *Herbarium*, where you re-created a false sense of scientific inquiry. And I wonder about this as it relates to a sense of truth in photography, which is something you've been very interested in for a long time.

JF: Yes, the problem of truth has been the main direction in my photographic work. One of my "bibles," or my cult books, has always been *Evidence*, a book published in 1977 by the Californian artists Mike Mandel and Larry Sultan, who collected photographs from different scientific institutions in the California area dealing with agriculture, genetics, forensic science, space research, etc. And they selected several pictures that were very informative and functional in their original context, but outside that original physical space—of the laboratory or research area—the pictures became absolutely surreal, very mysterious, and completely enigmatic because they were lacking all framework. I mean, you were lost in the search for meaning.

So there are these issues of ambiguity and truth, how we construct truth and how we construct meaning—all of which help us to understand the constellation of relations with the physical space, the institutional space, the political space, and so on.

For instance, right now I'm curating an exhibition at the museum of the University of Alicante, in Spain, titled Microcosmics, which is a sort of parody of your work as a biologist and an artist. I asked different artists from different disciplines—architecture, poetry, filmmaking, sculpture, photography, painting, and so on—to look through an electron microscope, which is usually limited to researchers and scientists, and react to that experience. And the result has been fascinating! The poet Carles Hac Mor, for example, didn't even look at an enlarged object, but at the microscope itself, at its entrails, which struck me as a way to take advantage of the "noise" produced by the microscope. I mean, I don't know how useful this may be, but as an artistic gesture, I think it's powerfully symbolic.

ARA: That sounds like a terrific exhibit. And this notion of "noise" that you mention is quite exciting, because one of the things that we constantly talk about in science is the signal-to-noise ratio, which is critical to determine the authenticity of a message. Noise is a critical part of meaning, much like silence.

JF: Let me tell you something that you will like. Recently I gave a lecture in Madrid introducing your work, the work of a German artist called Joachim Schmid, and my own work, and the title was "Photographing the White Noise." Schmid assembled tons of anonymous pictures and classified them according to archive criteria, like wedding, sports, travel, etc. Then he shredded all the photos in the same category, so he got millions of colorful thin threads. He then recomposed these in a collagelike image, which follows a horizontal-linear pattern. In the resulting picture you can find all the information which was originally in those pictures, but combined in a random way, so you get a sort of snapshot of static, like you have on the TV screen when you're not tuning it properly. The information hasn't been destroyed—everything's still there—but it's become absolutely useless. And the work is aptly titled *Statics*.

ARA: Yes, I think that's excellent. This reminds me that I wanted to discuss portraiture, and what makes a portrait successful. We started with the idea of landscape, and in a way a portrait is sort of a subdomain of landscape. I wonder how much the idea of portrait has changed, and to what degree portrait, or identity, can be changed by the way the information is transmitted and contained. This is the subject I have explored in curating the next issue of *Photovision*.

Also, reading your book, *The Kiss of Judas: Photography and Truth*, I've been struck by your interest in things forbidden. So I wonder to what extent the concept of portrait and the concept of landscape are intimately tied to the boundaries of power.

JF: Well, I believe that those categories are absolutely obsolete in contemporary image-making. I mean, we can talk about portraits of space and we can talk about landscapes of the face. So these are just academic genres or old-fashioned ways to categorize art. Within what we understand as landscape or what we understand as portrait, there are many, many new and novel approaches.

For instance, I recently discovered the work of the Argentinean artist Leandro Berra, who did a wonderful project dealing with identity in portraiture. Berra went into exile in the mid-seventies because of the Argentinean dictatorship, and he settled in Paris. And just very recently he found out that one of his best friends who had been "missing" was actually killed by the political police of Argentina's Junta Militar. And so Berra wanted to re-create his friend's face but he wasn't able to remember enough to do it well.

So he went to the police in Paris and asked for assistance—which was kind of ironic—and got access to a system used by Interpol and other police agencies to reconstruct the faces of missing persons, or criminals, and he tried to reconstruct the image of his friend. But he couldn't. It was impossible for him to approach what he felt was the true face of this person. There was always some unavoidable noise that created a gap between the real model and its artificial reconstruction.

But this failure gave him the idea to somehow use this "noise" in portraiture, noise from the process of creating a portrait. So his current project consists of asking volunteers to re-create their own faces without looking in the mirror or at a photograph—using only their memory. I did the experiment myself, and it was brutal because it's hours of trying to select the eyes, the best nose, the ears, hair, and so on—for each category you must decide among 1,600 possibilities. So it was really a compelling challenge because finally you realize that you don't know at all how to project your own ideas, your own clichés about your appearance. It's a wonderful experiment about essence and appearance, about substance and façade.

I think that such proposals make portraiture a category to keep alive despite the crisis in which I believe these traditional categories are right now.

ARA: But still you have a portrait as the final piece of identification. In some countries they ask for photographs in which you see an ear. I guess because the shape of an ear is quite peculiar. So there is still the idea that you can be identified visually by those in power. And that's still very much used, even through fingerprinting, DNA fingerprinting, and other methods with more biological criteria are extensive.

What you were saying reminded me of a process that I've tried to incorporate in other bodies of work—*Memory* and *Genome and Identity*—in which the issue is how we know or remember somebody. How do we remember a portrait, a face? Where does the knowledge for the whole, but not the parts, come from? Because we can certainly recognize people even from bits of images of their faces, so we're able to reconstruct a whole.

JF: Yes, well, that's the dialectic between the fragment and the whole. And this was explored by nineteenth-century ideologies on eugenics. I mean, Bertillon and Lombroso, with all these catalogs of different body fragments, meant to recombine and produce permutations of different types of people, and so on. Or Francis Galton, with his photosynthesis of different individuals, trying to create archetypes of thieves, murderers, homosexuals, etc.

I think this is an important aspect of questioning how we achieve an identity, how we recognize something.

ARA: But don't you think that there is a crisis of identity right now that is mirrored by the sort of deep will to find a "true" biological, molecular identifier?

JF: I feel that when we talk about identity, we talk about many dif-

Joan Fontcuberta and Ariel Ruiz i Altaba

ferent things. Identity doesn't mean only to exist, but to hold a certain uniqueness. I consider identity as a sort of validation of my roles, my attitudes, my circumstances. For instance, I'm an artist, or a husband, or a father, or Catalan.

ARA: You're saying identity is an operational definition. And how you define yourself depends on what you're doing, who you're interacting with, how you're applying yourself, etc.

JF: Exactly. For me it's a functional concept.

ARA: Right. It's quite wonderful, in a way, to think that there's no static definition because very often people find solace in the idea that there are rock-solid concepts that don't change. And this idea, in a way, is very dangerous for society, art, and science. Because it prevents fluidity of change and novelty.

JF: Yes.

ARA: Then do you think that the idea of truth, which you've focused on for some time, is also subject to change? Do you think of truth as not existing anymore, but being operationally defined in the moment something is happening?

JF: Yes. Absolutely. I think that truth is a convention. Truth is a sort of agreement, and depends on point of view. There's no objective truth. There's just a sort of agreement regarding how accurate the information is with respect to a model. So we should talk about systems of knowledge or models of knowledge rather than truth. Truth has become a poetic concept—

ARA: So you think science is based on a poetic concept?

JF: Why not? But I'm a little pissed off that science seems to have renounced this poetry that it had at its origin. Recently I had a bizarre experience: I visited the Instituto de Astrofísica in the Canary Islands. This is an astronomical observatory on the summit of the Teide volcano on Tenerife Island. It's one of the best in Europe and I was curious to observe through one of these professional, technologically very advanced telescopes. And I was absolutely disappointed because you don't see anything at all! I mean, all the information is projected onto a computer screen and everything is represented in numbers. So let's say that I feel that science is losing some poetic visibility. This abstraction is probably very useful, but it's without the glamour, the sensuality, that, you know, science used to have.

There have been a lot of exhibitions and books about the relationship of photography and science, but my favorite one is that compiled and edited by Ann Thomas—

ARA: In Canada, yes.

JF: Yes, the curator of the photography department at the National Gallery in Ottawa. It was a book and an exhibition called *Beauty of Another Order*, and all the pictures in that book, which had been done for scientific purposes, had this poetic and sensual dimension. And I feel they won't be able to do a similar book in the next twenty years because it's possible there won't be pictures, just numbers. I don't know if you agree with that.

ARA: Well, when you talk about the book *Evidence*, or this book curated by Thomas, I think that there is a very critical element to consider, which is selection. In both cases, these people selected through thousands of photographs to find those that fit their intended message. And therefore, I think it would be unfair to say that science is, or scientific images are, more or less poetic, then than now.

Also, I think it is critical to realize that science may be poetic, but if it were to remain only poetic, we would all be dead. Because it's very nice to say that science is poetic, but in the end somebody has to go beyond poetry and invent a new drug for the resistant bacteria that are going to infect your lungs, or somebody has to find a new cure for AIDS or cancer.

So I see both sides, perhaps in a schizophrenic way, which is exactly how I describe my endeavors in an introductory text that I wrote for a show titled "How We Look?" at Princeton University. So when I hear that science is losing a poetic visual realm, I would say that all images are messages. Before, many messages in science were conveyed through drawing, but there's now a tremendous pressure to have photographs rather than drawings, as if the photograph, and not the drawing, was giving you the truth of the message. But in any case, science is based on the idea that there *is* a truth, whether or not we have access to it. What we *know*—I agree with you—depends on systems of knowledge, and we may never be able to get to specific truths, but the idea is that there is *a* reality; there is a way humans evolved or cancers grow, for instance. And science is a way to get to that.

In terms of visibility, the science that becomes nonvisual is the science that is more evolved, and in the end, you get to the realm of mathematics. One of my brothers, Martí, who studies superstrings, can try to tell me how beautiful something is in five or eleven dimensions, but I just don't even grasp it. Now does that mean it's less poetic? I say no. I think the difference in its poetry is whether you can access it through knowledge of the language used.

JF: Yes, this problem of translation is exactly what I deal with in the *Orogenesis* project, which I mentioned. I confront landscapes and maps as translations of space experience.

ARA: So how do you see these issues of codes, of language, of structures of power, and the idea that science may be different?

JF: Well, that's one of my favorite subjects. I mean, in addition to science I've dealt with mass media, politics, religious beliefs, and art history. My work is trying to react to any kind of institutional platform of authoritarian discourse. So I think that science cannot present itself as having a monopoly on truth; that science is only a set of temporary, or provisional, truths, just waiting for a new interpretation or a new theory.

And, of course, the idea of power is crucial to the construction of meaning. And because of that, most of my work is being ironic, and creating parodies of a power situation—who is talking and who is listening. All these relationships establish different power situations.

ARA: Right.

JF: You said something that I wanted to pick up on—the fact that photography is providing more convincing information than drawing.

I mean, why is photography, which is rooted in the techno-scientific culture of the nineteenth century, considered a neutral transcription of reality, and a drawing considered a subjective interpretation? I believe this is because of the power of technology; I mean, the fact that technological images have that charisma, that prestige, that reputation in terms of providing accurate neutral depictions of the world. And this is just, of course, an ideological convention. I mean, there's nothing in the technological nature of photography that can ensure that the supposed objectivity exists at all. This is just a sort of myth. And I'm surprised that, in circles of intelligent people like scientists, it still has the kind of acceptance it does.

ARA: Oh, absolutely. This question is one of the reasons I started working on images before I started working on science. And I think I've gravitated toward a science that relies very heavily on images. Because I guess what I'm most interested in, at the end of the day, is, What is in an image? What's the information encoded in that image; and what's the power of that image?

We see that the manipulation of an image's meaning depends on context, as you were saying before. If an article of science is published in a high-quality, heavy-paper, shiny, beautifully set up journal, immediately those images have more power. Their reproductive fitness, let's say, is high.

JF: Right.

ARA: One other issue I'd like to pick up from what you were saying is the reality of photographs. My other brother, Cristián, is an evolutionary biologist, and we often discuss whether there is a problem when museums use photographs to substitute the object. Viewing a photograph gives the sense that the object exists, even when it doesn't.

For example, in the newspaper recently, there was a photograph of a lemurlike creature discovered in the forests of Borneo, which was claimed to be a "new" mammal. The only thing you saw were two shiny eyes reflecting the light of the flash, and the outline of a body with a big, long tail. And this tells us it exists. Fine. At the same time, the caption says it is in danger of being extinct. So what is the real message?

There is a fine balance between seeing, knowing, and being, and we're living an unprecedented disaster of our own making, which is the destruction of biodiversity, of the natural world, which, interestingly, correlates precisely with an enormous explosion in digital imagery.

And so we begin to live in a virtual world, where as long as we see images of things, we don't question whether they actually exist.
JF: Yes. I would say that any use of photography is dangerous, not only in museums. I mean, it ventures into schools, it ventures into the media, and it even ventures into family albums. But we should decide which kind of danger, which kind of risk we are talking about—and what kind of risk we can accept in exchange for its benefits.

I always think of a museum—I think it was in Philadelphia—that was displaying a diorama of dinosaurs, with painted landscapes and animals and so on. And I noticed that there was a small caption, which stated when that information was updated.

And I thought that this was proof of modesty, of a humble attitude on the part of that paleontologist who was aware that knowledge was limited and would change. Of course, in that case it was not photographs. They were just paintings and models. But I think photographs in museums should include the same warnings or caveats.
ARA: Right—to update the intended message. Images are too power-

ful to be ignored or to let them creep into our systems of knowledge without an inquiry.

JF: This reminds me about something we did in my book *Contranatura*. As a joke, I inserted a fake advertising page selling the concept of Critical Skepticism, with the symbol of registered trademark and so on. The motto was "Critical skeptics of all countries, unite!" ∞

Joan Fontcuberta and Ariel Ruiz i Altaba

Six

Climate Politics
Laurie David and Stephen Schneider

The activist and climate scientist deliberate the state of the planet.

When Laurie David began recruiting scientists to appear in a documentary on global warming, she remembered Stephen Schneider's informed and articulate comments in the wake of hurricane Katrina, and immediately invited him to participate. The film, *Too Hot Not to Handle*, aired on HBO on Earth Day, April 22, 2006, and features Schneider, among other experts. When they caught up soon after to deliberate the state of the planet, Schneider and David had no shortage of fodder for discussion, from Michael Crichton to Al Gore, Greenland to Detroit, CNN, SUVs, the GDP, and latkes.

Laurie David is an author, a global-warming activist, and a producer of the Academy Award–winning film *An Inconvenient Truth*, the HBO documentary *Too Hot Not to Handle*, and the prime-time comedy special *Earth to America!* A trustee of the Natural Resources Defense Council, David founded the Stop Global Warming Virtual March with Senator John McCain and Robert F. Kennedy Jr. and authored the best-selling *Stop Global Warming: The Solution Is You!* and *The Down-to-Earth Guide to Global Warming* for children. Called the Bono of climate change, David has been honored by the National Audubon Society, the National Wildlife Federation, and the U.S. Environmental Protection Agency, receiving its Climate Protection Award in 2008.

Stephen Schneider is the Melvin and Joan Lane Professor for Interdisciplinary Environmental Studies at Stanford University. He received his PhD from Columbia University and studied as a postdoctoral fellow at NASA's Goddard Institute for Space Studies. Schneider has served under seven presidential administrations, including the current one, as either a consultant to federal agencies or as a member of the White House staff. The author and editor of numerous books, he received a MacArthur Fellowship for his efforts to increase public understanding of climate research. He also shares a Nobel Prize with his colleagues on the Intergovernmental Panel on Climate Change and Al Gore. His latest book, *Science as a Contact Sport*, is a behind-the-scenes account of the battle against climate change.

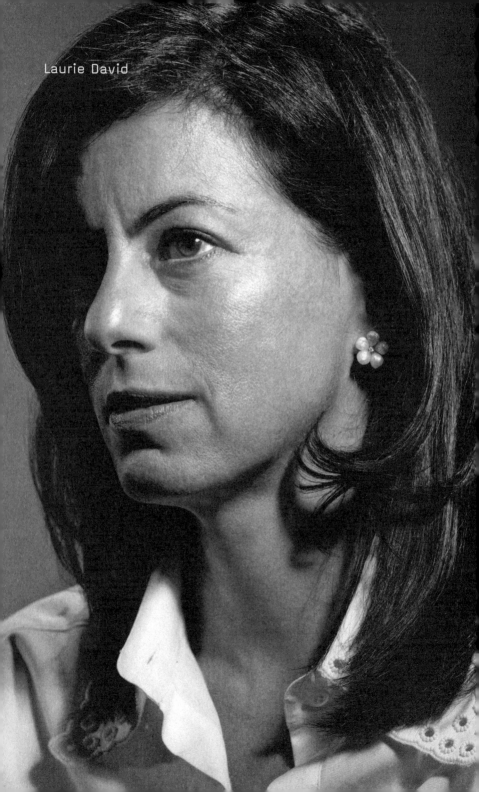
Laurie David

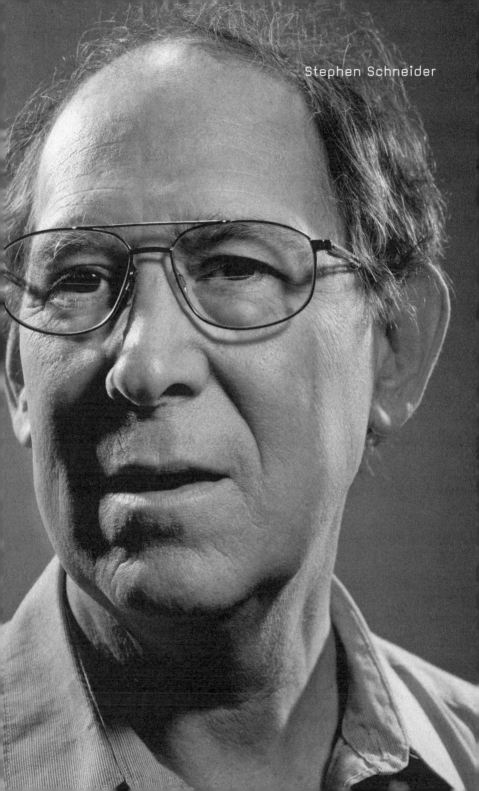
Stephen Schneider

Laurie David: I want to start with an obvious question: Does the truth matter? How are we going to educate the public if they're not told the truth about the science from our scientists? Obviously, what I'm talking about is what happened recently with James Hansen and NASA.

Stephen Schneider: My students are always saying, "Aren't you frustrated to death? Nothing you do makes any immediate difference." What I keep trying to tell them is, the truth matters, but it's on a generational time frame. In the short run, it's all political spin: media, chicanery, and who buys the airwaves. But in the long run, being right and having events occur the way you said they would builds credibility, and then some phenomena comes along and becomes a tipping point. In 1988 it was the super heat waves, which tipped the global-warming problem—as I like to say—from the left brain of one hundred of us to the right brain of the society. And that of course set up the Global Climate Coalition—the coalition of liars and spin doctors and others who then spent tens of millions of dollars a year repositioning the debate. And for a decade they were successful.

Now we have another tipping phenomena in Katrina. Nature is cooperating with theory. We continue to break warming records and hurricane intensities are increasing in correlation with warming oceans, exactly as predicted by theory fifteen to twenty years ago. Slowly, that works its way through—although in fits and starts from these media-worthy events—so that after a generation or two, when problems become pretty widely understood, the truth matters. But in the short run it is going to be all spin.

LD: Well, the problem is we don't have any short run left, right? Everyone's saying that we've got ten years to substantially slow global warming down. Ten years. So we better get moving.

SS: I would argue that we're way past having ten years to avoid some damage that we just absolutely can't prevent. We still have time to avoid the really large, ugly, horrible stuff, the Gulf Stream flipping off, the melting of—

LD: Of Greenland.

SS: The west Antarctic and Greenland ice sheets. But we should have started much more aggressively twenty-five years ago when we—I'll say patting me, Jim Hansen, and Mike Oppenheimer on the back—kept arguing, and saying, "Let's slow this down now. Let's do it in an orderly manner and let's do it cost effectively" and all we got was, "Oh, you're doomsters and gloomsters and this isn't proven," and they put us on a path that's not sustainable. So now we have to take more action, that's more costly, and the more we delay, the more expensive it's going to be and the more climate change we'll have to adapt to. It's not like the Earth turns into a climatic pumpkin above two degrees warming and below that everything is fine. It's the continuous adding-up of systems at risk.

LD: Right, but I would argue—and you're likely going to disagree with this—that scientists are the most cautious people on the planet. And the time for caution on this issue has long passed.

SS: Well, it all depends on who you're talking about. It's true, most scientists are very uncomfortable saying anything they're not absolutely sure of, but that's certainly not true for the Mike Oppenheimers, or me, or others.

LD: No, there are a handful of you out there.

SS: You dial me on Google and ten thousand sites will say I'm an exaggerating liar because I've long believed—for more than thirty years—that when you see the planetary life support system getting messed up, you don't wait for full 99 percent certainty. You slow it down. We buy fire insurance when there's less than a 1 percent chance our house is going to burn down. We have a military, and although I may not like everything we do with it, I don't know anybody who says you should get rid of it because you have security precautions against only very low probability—but potentially dangerous—threats. Well, the climate change threat is not 1 percent. It's better than 50 percent for really significant trouble, and maybe 10 percent for absolutely catastrophic trouble. What kind of crazy person would take that chance when you can fix it relatively easily? By which I mean below the growth rate of the GDP.

LD: Right, right.

SS: I don't mean it's going to be cheap or politically simple.

LD: You gave a great analogy in *Too Hot Not to Handle* about global warming—

SS: Oh, yeah. I actually stole that from Richard Somerville. He said if you have high cholesterol and the doctor says, "Hey, you better stop eating that fatty food, you better get on Lipitor," do you turn around and say, "Oh, no-no. When is the heart attack? How serious will it be? When will I have the first warning sign that I have to do something?" No, you take precautions, because you can't have that kind of detailed information. We have very clear, overwhelming evidence that it is warmer. And we have virtually overwhelming evidence that humans are at least half the story. My own personal view is that we're going to become 95 percent of the story over this century. That's just way, way too high a preponderance of evidence to ignore.

LD: Part of what I work on is trying to get this story off of the science pages and onto the front pages and into popular culture. I think it's been marginalized because it's been viewed solely as an environmental issue, which I don't believe it is anymore. I view global warming as a huge national security threat, a public health crisis, an economic crisis, and I think part of what we have to start doing is figuring out a different way to talk to the American people about it so that they start paying attention.

SS: As I keep saying, the truth is bad enough. We don't have to overstate the case. We don't need *The Day After Tomorrow*. And what happens when we do overstate it is we lose our credibility and we're back in the "he said, she said" debate. It plays right into the hands of the lazy people in the media and next thing you know you've got a scientific phony like Michael Crichton on *Oprah*. I've got to bust my butt to get on *Science Friday* and we needed Katrina to get me on *Real Time with Bill Maher*.

LD: What Michael Crichton did was criminal because he promoted a book of pure fiction as if it were factual. And for some Americans, that will be the only thing they read about global warming.

SS: It's unbelievable. And this idea that, in science, if you're not absolutely certain then it's "just a theory" was pushed by the Global Climate Coalition and the Bush administration. We didn't have 99 percent certainty that there were weapons of mass destruction in Iraq, yet this government is a big believer in precautionary insurance against potential threats to security. But not when the security threat is rising sea levels or forest fires in California. Oh, for that we need to be 99 percent objectively sure. It's dripping with hypocrisy and it has nothing to do with the misunderstanding of science. That's what a lot of scientists don't understand. We're not miscommunicating our science. We're running into special interests who will lie on a dime for multimillion-dollar clients, and they'll do it continuously in order to maintain market share and delay action. And scientists are blindsided by it and don't know how to deal with it.

LD: You're completely right, but what's astonishing then, is the fact that this year—and I am totally feeling this—we're winning the general debate that the globe is warming and humans are causing it. We're winning that despite all the efforts of special interests to misinform the public or to create confusion about this issue.

SS: Yeah, I'm watching the moderate Republicans in Congress and even some of the "fossil fuel Democrats" begin to move toward the center and argue that we need national policy on this. And the large companies are starting to recognize—just like the chemical companies did when the ozone hole was discovered in the 1980s—that fighting this with full-page ads in the *New York Times* and vilifying the scientists is not really a winning strategy over the long-term. Because they're flying in the face of the truth, in the face of history, and they know eventually it's going to catch up to them. I think we're starting to see that in the energy industry, and in the auto industry with production of hybrids. We're just beginning to make progress.

That's the good news. The bad news is people still want to blame somebody else. They're not thinking about that gas-guzzling clunker they've got outside their door as part of the problem.

LD: And the thing is, this isn't about sacrifice, it's only about change. I drive a hybrid car, and I've been driving it for years and it gets me everywhere I need to go, in style and on time. The things that we need to start doing are really about a change of attitude.

SS: And I have no problem with you having a utility vehicle if it's for utility. But we don't need to drive them to the supermarket in California. It's not only bad for the environment, it's also bad for our own national security because it puts subtle, if not overt, pressure on us to use blood and treasure to defend the lines of oil supply, and to start moving into fragile offshore zones and into wildlife refuges. We could eliminate that need by just driving sensible automobiles.

LD: We just have to get Detroit to make them.

SS: Well, if Detroit doesn't make them, I'm very happy to help put them out of business.

And right now gasoline is cheaper than mineral water. That's why we have those monsters on the road, because gas prices are way too low. And when I say that, immediately people say, "Oh, you're going to hurt the economy."

LD: Nothing is going to hurt the economy more than global warming. That's clear.

SS: But there are solutions, too. If we increase the price of gas, for

Laurie David and Stephen Schneider

example, let's have a transition period for poor people driving old cars that have bad mileage. We need a voucher system so they can buy efficient cars, and the vouchers can be in proportion to mileage. And they can buy in Detroit—they'll help Detroit make money. We have to have creative ideas like this to connect development and the environment instead of pit them against one another.

LD: Wouldn't it be amazing if Ford offered a program to buy back dirty old cars?

SS: There you go, right. Some churches used to pay people to bring in their handguns. A buy-back program. They'd get such good publicity for it.

LD: They'd get amazing publicity and they'd be doing something positive, something real.

SS: What an ad campaign. They could take the bought-back cars that are old and clunky and not send them to Mexico, but follow them over to the junkyard where they get compressed, and then back to the steel furnaces where they get recycled. They could follow the whole life cycle of the car. People would be impressed. People are pretty impressed by decent behavior.

LD: Absolutely. You're totally right. But instead the Alliance of Automobile Manufacturers is suing the state of California for trying to curb global-warming emissions.

SS: It's outrageous. I was talking to one of the Japanese competitors and they said, "What do we need to do to take over the environmental market in California?" And I said, "Produce a plug-in hybrid and then publicly drop out of the lawsuit against the Pavley bill." I said, "This *is* the field of dreams. If you build it, they will buy it, but you've got to build it first."

LD: Right. Exactly.

SS: They told me that would be aggressive. I said, "You asked me how your company is going to take the number one environmental position. Well you've got to stand for something."

LD: Will we ever have a politician who is brave enough to fight for a gas tax?

SS: Well, Al Gore used to be. And he's back there again.

LD: Yes, he is. I actually just produced a movie called *An Inconvenient Truth* about his thirty-year journey studying and working on global warming. And I picked up *Earth in the Balance*, which I'd actually never read at the time he wrote it, fifteen years ago, and I tell you every word in that book has come true.

SS: I said earlier that events matter, but there has to be somebody out there who can turn that event around and make the connection. And Gore seems capable of that.

LD: Making the connections is so important, which is why I want to talk about the weather. This winter is breaking every temperature record. There are people golfing in Maine in January.

SS: Watch out, they're going to like this.

LD: See, that's the argument that makes me crazy. That people say, "Well, what's one degree more, or two degrees, okay so the sap will come earlier." Nobody can see past their nose to how everything is related and what the domino effect is of this warming.

SS: Well, you can never point to short-term warming or cooling for anything.

LD: I know, but we have an opportunity here to get the average person to connect the dots. This is the one thing that they are personally living through—this bizarre weather. I know you can't base overall climate on any one day of weather, but when, month after month, this is happening, we're making a mistake if we don't connect the dots for people. We're still not at a place where one newscaster ever says the words "global warming" when talking about these severe weather patterns.

SS: After Katrina and then Rita, I tried to do it, every time I spoke to the media. I said this isn't only nature at work here. No, we don't make hurricanes, but we can soup them up. As Bill Maher asked when I did his program, "Did we put our hurricanes on steroids?" It's not a bad metaphor. But it's very difficult, because NOAA—the National Oceanic and Atmospheric Administration—says we don't have enough information to know whether humans are impacting hurricanes. So they select information on the frequency, which is equivocal, and they entirely ignore all the new studies on the intensity, which is what causes damage. One Category 5 is way worse than ten Category 2s.

LD: But that's another way the public is lied to, when they only get one piece of the story.

SS: That's right, and when they leave that out, that's not accidental; that's a lie. Because if you know there's another component to the story, well established scientifically, and you deliberately leave it out, you're misleading on purpose, which is a violation of every principle a scientist has in our unwritten Hippocratic oath.

LD: I saw Wolf Blitzer interview Bill Clinton on CNN, and they edited what Clinton said. He said, "It's not causing more frequent hurricanes, but it's causing them to be more intense." And they cut the part, " . . . but it's causing them to be more intense."

SS: It's reprehensible. I'd fire a journalist for doing that. Or the editor. Because that's sanctioning lying. It's lying. There's no other way to phrase it.

LD: I totally agree. Let me switch gears with you for a second. I want to ask you, because you're a professor, how do we engage students on this issue? One of my big disappointments right now is with the student body.

SS: I'm probably a biased observer because I don't see that with my students at all. They're not only concerned, they're working and volunteering for environmental groups in Washington and for members of Congress, and learning everything they can so that they can be more effective. I can't wait for my kids to take over running this world, because my generation has screwed it up so royally with a combination of ignorance, greed, xenophobia, and religious zealotry, and I don't see most of that in my kids.

LD: But they have to start making more noise. We have to hear from them.

SS: I was a grad student in '68 and I remember Bob Dylan changed the culture. He said something like, "Patriotism is not listening to the president, it's following the Constitution." That was a radical concept then. Now it's not, so there isn't quite that degree of foment—I don't see that there's less passion and concern for a sustainable future now than then. Because it's not so radical or as rebellious as it was back then, maybe we're not attracting the kind of people who were stomping out on the streets. But they're certainly working through established channels to make a difference. These kids are not sitting on their hands.

LD: One of the things I did last year was start a yearlong virtual march to stop global warming.

SS: Yeah, I've gotten quite a few miles on it.

LD: Yes, you were one of my first marchers. And we need everybody to go to the Web site and sign up to march, virtually, and become part of this, because the goal is to make the numbers so big that the media and Congress and the administration can't ignore this anymore. Basically we're saying the globe is warming, people are causing it, and we want meaningful solutions now. And marching in the streets, unfortunately, doesn't really work anymore. It ends up being thirty seconds on the evening news.

SS: And then they'll find the seven marchers from the rapturist society—

LD: And they'll give them equal coverage.

SS: And then they'll get Michael Crichton to say why scientists are elitists and they don't believe in democracy. What are you going to do? You have a media that is derelict and lazy. They use the excuse that it's balanced if you get the Democrat and get the Republican.

Well, in political reporting that's balanced, but not in science. In science you have to report the quality of the arguments, not simply who said them.

LD: Exactly. In their goal of striking balance, they strike complete and total imbalance.

SS: Let's go back to Bush for a second, just to show you the horrible misrepresentation in the popular debate about global warming. Back in the early nineties, I challenged some economists who said it was too expensive to have climate policy. What they were arguing was that we would be 500 percent richer in 2105 with climate policy that eliminates most of the severe global warming rather than 500 percent richer in 2100 without it. This is crazy. It's a cheap insurance policy. What they do is—I call it a Carl Sagan problem—they talk about the "billions and trillions" that have to be spent. What they forget to tell you is that that's a small fraction of the overall growth rate of the economy as projected into the century. And last summer Bush said that if we'd signed on to the Kyoto Protocol that it would have bankrupt America. Either he's ignorant or he's lying. There is no option in between.

LD: Right, yeah.

SS: It was completely untrue. And in fact that's been proven, because gasoline in the United States went up over a dollar a gallon in 2005. And it wasn't due to environmental rules. That's equivalent to a tax on carbon of over more than $200 a ton. The most pessimistic economists estimated the cost of Kyoto to the U.S. economy at $100 to $200 per ton of carbon. So we had a situation—without any environmental policy—that cost more than what Kyoto would have cost. And what did it do to the economy? Next to nothing, just like the models said.

So Bush was sitting there, right in the middle of the countervailing proof, making a ridiculous statement, and the media never challenged it. It was unbelievable. I told everybody I saw why that was crap. His silly statement just sat there, like a latke.

So I don't know—how do you get anybody to listen, if nobody tells the story straight?

LD: Okay—who else but you would refer to a latke? Well, you know what, that's what we're working on. And we're going to do it. I think first you have to get your message out there, start blogging.

SS: Well, my Web site, as you know, is three hundred pages long and not written for the average congressional staffer or for the media. It's written for the people who want to have some in-depth capacity to defend their positions and win a debate; that's where I've been putting my energy.

LD: I know, but there's a power to unlikely messengers in unlikely places. And look what we've accomplished this year. Oprah Winfrey did her first show—in twenty years of television—on an environmental issue, and she did it on global warming.

SS: Yeah, Mike Oppenheimer was on that, right?

LD: Right. Mike did it, and Leonardo DiCaprio, and I was on it. That was a major step forward. This year Fox News did an hour-long special on global warming—in prime time. You didn't see that on NBC or CBS or ABC—

SS: Yes, I know. They shocked me when they actually had a story during Katrina that was accurate. I was amazed. I hope it's a trend.

LD: Yeah. People couldn't believe it. So those are two big prime-time hits, not including ours on HBO this month. So we've got to keep plowing ahead. We're all part of this problem and we all need to be part of the solution.

Stephen, I want you to try to use the latke reference again before we're finished.

SS: I'm glad to see it didn't leave you flat. ∞

Seven

War and Deceit
Noam Chomsky and Robert Trivers

The antiwar activist and MIT linguist meets the Rutgers evolutionary biologist to discuss war and the psychology of deceit.

In the 1970s, a Harvard class taught by Robert Trivers ignited a controversy that would escalate into the "sociobiology wars." Across town at MIT, Noam Chomsky had earned a reputation as a leading opponent of the Vietnam War. Throughout those pivotal years, and in the following decades, the two explored similar ideas, although from different perspectives. Long aware of each other's work, they had never met until 2006, when they sat down to compare notes on some common interests: deceit and self-deception.

Noam Chomsky has written and lectured widely on linguistics, philosophy, intellectual history, contemporary issues, international affairs, and U.S. foreign policy. His pioneering work in syntactic structure revolutionized linguistics and has influenced fields as diverse as cognitive psychology, computer science, and philosophy. Chomsky is the recipient of numerous honorary degrees and awards, and has been called the world's most important intellectual by the *New York Times*. In 1955 Chomsky joined the staff at MIT, where he continues to work as a professor of linguistics.

Robert Trivers is an evolutionary biologist. In the early 1970s he wrote a series of landmark papers that provided a Darwinian basis for understanding complex human activities and relationships. He was awarded the prestigious Crafoord Prize in Biosciences by the Royal Swedish Academy in 2007, for his "fundamental analysis of social evolution, conflict, and cooperation." Trivers's most recent book, *Genes in Conflict* (with Austin Burt) presents the first comprehensive review of selfish genetic elements in all species. He currently teaches at Rutgers University.

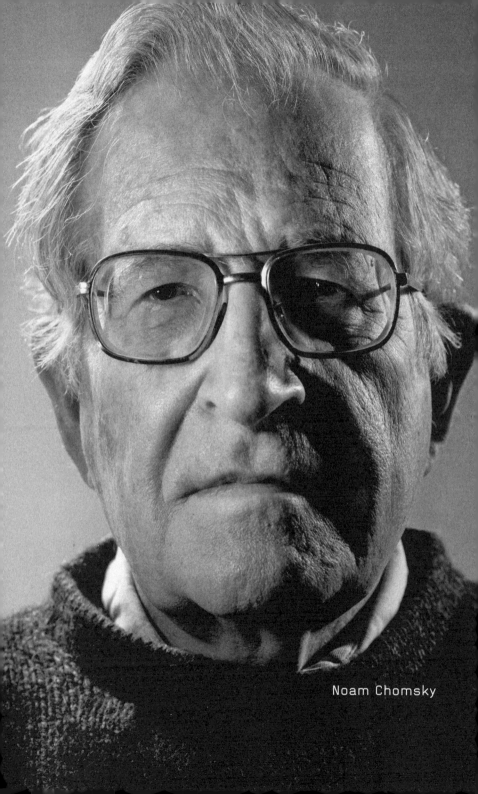

Noam Chomsky

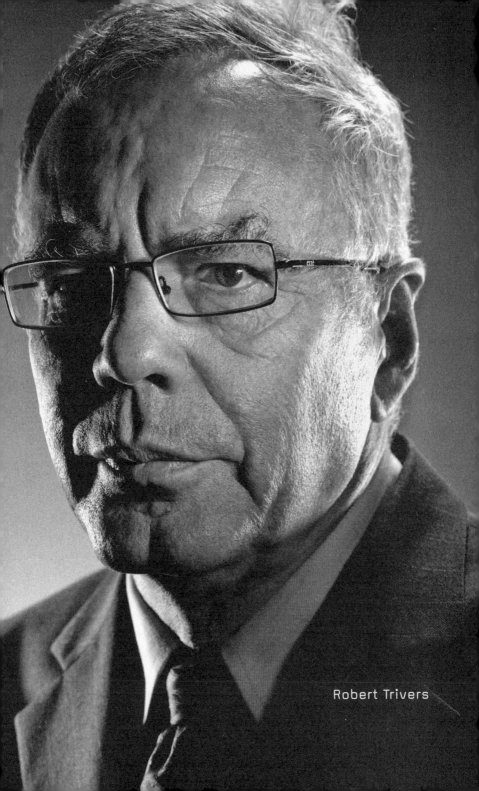
Robert Trivers

Noam Chomsky: One of the most important comments on deceit, I think, was made by Adam Smith. He pointed out that a major goal of business is to deceive and oppress the public.

And one of the striking features of the modern period is the institutionalization of that process, so that we now have huge industries deceiving the public—and they're very conscious about it, the public relations industry. Interestingly, this developed in the freest countries—in Britain and the United States—roughly around the time of World War I, when it was recognized that enough freedom had been won that people could no longer be controlled by force. So modes of deception and manipulation had to be developed in order to keep them under control.

And by now these are huge industries. They not only dominate marketing of commodities, but they also control the political system. As anyone who watches a U.S. election knows, it's marketing. It's the same techniques that are used to market toothpaste.

And, of course, there are power systems in place to facilitate this. Throughout history it's been mostly the property holders or the educated classes who've tended to support power systems. And that's a large part of what I think education is—it's a form of indoctrination. You have to reconstruct a picture of the world in order to be conducive to the interests and concerns of the educated classes, and this involves a lot of self-deceit.

Robert Trivers: So you're talking about self-deception in at least two contexts. One is intellectuals who, in a sense, go through a process of education that results in a self-deceived organism who is really working to serve the interests of the privileged few without necessarily being conscious of it at all.

The other thing is these massive industries of persuasion and deception, which, one can conceptualize, are also inducing a form of either ignorance or self-deception in listeners, where they come to believe that they know the truth when in fact they're just being manipulated.

So let me ask you, when you think about the leaders—let's say the present set of organisms that launched this dreadful Iraq misadventure—how important is their level of self-deception? We know they launched the whole thing in a swarm of lies, the evidence for which is too overwhelming to even need to be referred to now. My view is that their deception leads to self-deception very easily.

NC: I agree, though I'm not sure they launched it with lies, and it's perfectly possible they believed it.

RT: Yes.

NC: I mean, they had a goal—we don't have a detailed record, but from the record we have, it's as if they sort of cherry-picked and coerced intelligence to yield evidence that would contribute to that goal.

RT: Yes.

NC: And anything that conflicted with it was just tossed out. In fact, people were tossed out—like the head of the Joint Chiefs of Staff.

RT: Right, indeed.

NC: I mean, I think we all know from personal life, if there's something you want to do, it's really easy to convince yourself it's right and just. You put away evidence that shows that's not true.

So it's self-deception but it's automatic, and it requires significant effort and energy to try to see yourself from a distance. It's hard to do.

RT: Oh, it is. I think in everyday life we're aware of the fact that when we're watching something onstage, so to speak, we have a better view than the actors on the stage have. If you can see events laterally, you can say, my god, they're doing this and they're doing that. But if you're embedded in that network it's much harder.

NC: In fact, you can see it very clearly by just comparing historical events that are similar. They're never identical, but similar.

Take the Russian invasion of Afghanistan, Saddam's invasion of Kuwait, and the U.S. invasion of Iraq—just take those three. From the point of view of the people who perpetrated these acts, they were each a noble effort and done for the benefit of everyone—in fact, the self-justifications are kind of similar. It almost translates. But we can't see it in ourselves; we can see it only in them, you know. Nobody doubts that the Russians committed aggression, that Saddam Hussein committed aggression, but with regard to ourselves, it's impossible.

I've reviewed a lot of the literature on this, and it's close to universal. We just cannot adopt toward ourselves the same attitudes that we adopt easily and, in fact, reflexively, when others commit crimes. No matter how strong the evidence.

RT: Not the overall crime.

NC: In fact, here's another case like Afghanistan and Kuwait. Dick Cheney was recently somewhere in Central Asia—Kazakhstan, I believe. He was getting them to make sure to direct their pipelines to the West, so the United States can control them.

And he said control over pipelines is a means—these are tools for intimidation and blackmail. He was talking about if somebody else controlled them. Like if China controlled the pipelines, it's a tool of intimidation and blackmail. But if the United States controls the pipelines, it's just benevolent and free and wonderful.

RT: Yes.

NC: And I was interested to see if anybody is going to comment on that. I mean, as long as he's talking about somebody else's control, then it's intimidation and blackmail. The very moment he's trying to get them to give us control, that's liberation and freedom. To be able to live with those contradictions in your mind really does take a good education. And it's true in case after case.

RT: It's the psychology of deceit and self-deception. When you start talking about groups, there are some very interesting analogies. Psychologists have shown that people make these verbal switches when they're in a we/they situation, in a your-group-versus-another situation.

NC: Groups that are simply set up for the experiment, you mean?

RT: It can be. You can also do it experimentally, or you can be talking about them and their group versus someone who's not a member of their group.

But you have the following kinds of verbal things that people do, apparently quite unconsciously. If you're a member of my group and you do something good, I make a general statement: "Noam Chomsky is an excellent person." Now if you do something bad, I give a particular statement, "Noam Chomsky stepped on my toe."

But it's exactly reversed if you're not a member of my group. If you're not a member of my group and you do something good I say, "Noam Chomsky gave me directions to MIT." But if he steps on my toe I say, "He's a lousy organism" or "He's an inconsiderate person."

So we generalize positively to ourselves, particularize negative and reverse it when we're talking about other people.

NC: Sounds like normal propaganda. Islamic people are all fascists. The Irish are all crooks.

RT: Yes, exactly. Generalize a negative characteristic in the other. Another thing that comes to mind with respect to the Iraq case: There's evidence suggesting that when you're contemplating something—whether or not to marry Suzy, for instance—you're in a deliberative stage. And you are considering options more or less rationally.

Now, once you decide to go with Suzy, you're in the instrumental phase; you don't want to hear about the negative side. Your mood goes up, and you delete all the negative stuff and you're just, "Suzy's the one."

One thing that's very striking about this Iraq disaster is there was no deliberative stage, unless you—

NC: Go back a few years.

RT: Yes, unless you refer to the nineties, when there were a couple of position papers by these same groups that said, "Let's not go to war."

But once 9/11 occurred, we know that within days, within hours, they were settling on Iraq and they went into the instrumental phase in a very major way. They didn't want to hear anything of the downside.

NC: That was dismissed.

RT: It was dismissed entirely. And these firewalls were set up so there was no communication. And if someone came into Rumsfeld's office and said, "Well, gee . . ." Well, [General] Shinseki got an early retirement plan.

And Wolfowitz comes in the very next day and says, "Hard to imagine that we'd need more troops to occupy than to knock over." But that was established military doctrine; we'd known that for more than fifty years.

NC: Just didn't want to hear it.

RT: Didn't want to hear it. So, I'm trying to understand these phenomena at the individual level and also put them together in groups, since at times institutions act like individuals in the way they practice internal self-deception.

This was Richard Feynman's famous analysis of NASA and the *Challenger* disaster. I don't know if you ever read the analysis—it was beautiful. He said that, when we decided to go to the moon in the sixties, there was no disagreement in the society, for better or worse. Everybody said, "Let's beat the Soviets to the moon." And the thing was built rationally from the ground up, and by god, before the decade was out we were on the moon and back safely.

Now, they had a $5 billion bureaucracy with nothing to do. So they had to come up with rationales for what they did. So they decided on manned flight because it's more expensive, and they decided on the reusable shuttle, which turned out to be more expensive than if you just used a new shuttle every time. But they always had to sell this thing as making sense.

So Feynman argued that the NASA higher-ups were busy selling

Noam Chomsky and Robert Trivers

this pile of you-know-what to the general public. They didn't want to hear anything negative from the people down below. This was his analysis for how they came up with this O-ring nonsense.

They had a safety unit that was supposed to be involved in safety, but ended up being subverted in function, to rationalize nonsafety. And the classic example is, there were twenty-four flights, I think it was twenty-four, prior to the disaster. And of those, seven suffered O-ring damage—

NC: Detected.

RT: Detected, yes. In one of them the O-ring had been burned through—a third of the way through. Now, how did they handle this? It was statistically significant. They said, "Seventeen flights had no damage, so they're irrelevant"—and they excluded them. Seven had damage—sometimes at high temperature—therefore temperature was irrelevant.

Then they came up with real absurdities. They said we built in a threefold safety factor. That's to say, it burned only a third of the way through. But that, as you know, is a perversion of language. By law you are required to build elevators with an eleven-fold safety factor, which means you pack it full of people, run it up and down, there is no damage to your equipment. Now you make it eleven times as strong.

NC: And all of this data was available.

RT: All of it was available.

RT: There's an analogy here to individual self-deception. Information is often somewhere in the organism; it's just well-hidden. It's well down in the unconscious. And it's often inaccessible because you build up firewalls against it.

NC: Are there any animal analogs to this?

RT: Well, I don't know. I believe that self-deception has evolved in two situations at least in other creatures, and that it can be studied. I've suggested a way to do it, but so far nobody's done it.

For example, when you make an evaluation of another animal in a combat situation—let's say male/male conflict—the other organism's sense of self-confidence is a relevant factor in your evaluation.

NC: And that's shown by its behavior.

RT: Exactly—through its suppressing signs of fear and not giving anything away, and so forth. So you can imagine selection for overconfidence—

NC: For showing overconfidence, even if it's not real.

RT: Yes. Likewise, in situations of courtship, where females are evaluating males. Again, the organism's sense of self is relevant. We all know that low self-esteem is a sexual romantic turnoff.

So again, you can have selection—without language it seems to me—for biased kinds of information flow within the organism in order to keep up a false front.

NC: And it may be that the animal that's putting up a false front knows it's a false front.

RT: Yes, but it may benefit from not knowing—

NC: Because it's easier.

RT: Easier to do it and perhaps more convincing because you're not giving away evidence.

NC: Secondary signs.

RT: Exactly.

NC: Is there any evidence about that, or is it just speculation?

RT: What you heard is rank speculation.

NC: Can it be investigated?

RT: I do not know of anybody who is doing it on self-deception. There is excellent work being done on deception in other creatures.

To give you just one line of work that's of some interest: We find repeatedly now—in wasps, in birds, and in monkeys—that when organisms realize they're being deceived, they get pissed off. And they often attack the deceiver. Especially if the deceiver is overrepresenting him- or herself. If you're underrepresenting and showing yourself as having less dominance than you really have, you're not attacked. And the ones that do attack you are precisely those whose dominance status you are attempting to expropriate or mimic.

It's very interesting and it suggests some of the dynamic in which fear of being detected while deceiving can be a secondary signal, precisely because if you are detected, you may get your butt kicked or get chased out.

NC: There's a name for that in the international affairs literature; it's called maintaining credibility. You have to carry out violent acts to maintain credibility, even if the issue is insignificant.

RT: Right.

NC: It's kind of like the Mafia.

RT: Yes, I know, I've heard that rationale used for odious stuff— we're maintaining credibility, maintaining street cred.

NC: That's a common theme. It's usually masked in some sense, so it's the credibility of the West, or the Free World, or something or other.

NC: Are there ways of studying self-deception?

RT: Yes, there was a brilliant study by [Ruben] Gur and [Harold] Sackeim, about twenty years ago—which was a very difficult one to do then, you could do it much more easily now—based on the fact that we respond to hearing our own voice with greater arousal than we do to hearing another human's voice. In both cases we show physiological arousal—galvanic skin response is one such measure. There's twice as big a jump if you hear your own voice.

Now, what you can do is have people matched for age and sex, read the same boring paragraph from Thomas Kuhn's "Structure of Scientific Revolution," chop it up into two-, four-, six-, and twelve-second segments, and create a master tape where some of the time they're hearing their own voice, but a lot of the time they're not.

Then they've got to press a button indicating if they think it's their own and a second button to indicate how sure they are. But meanwhile they have the galvanic skin response.

Now they discovered two interesting things. First of all, some people denied their own voice some of the time, but the skin always had it right. Some projected their own voice some of the time, but the skin always had it right.

The deniers denied the denial, but half the time, the projectors were willing to admit afterward that they thought they'd made the mistake of projection.

NC: What do you think the reason for that is?

RT: The difference between the projectors and the deniers? Well, I don't have a good way of putting it, Noam, but to me when you want to deny reality, you've got to act quickly and get it out of sight. The deniers also showed the highest levels of galvanic skin response to all stimuli. It's like they were primed to do it. And inventing reality is a little bit more of a relaxed enterprise I suppose.

NC: It's not as threatening.

RT: Yes, something like that. The final thing Gur and Sackeim showed was that they could manipulate it. Psychologists have lots of devices for making you feel bad about yourself, and one of them is just to give you an exam. They did this with university students. Then they told half of them, you did lousy, and half of them, you did well.

And what they found was that those who were made to feel badly about themselves started to deny their own voice more, while those who were feeling good started hearing themselves talking when they weren't. Now since we didn't evolve to hear our voice on a tape recorder, we have to interpret here. But it's like self-presentation is contracting on your failure and expanding on your success.

But back to your question. Among animals, birds in particular have been shown to have the same physiological arousal that humans do—arousal to their own species song, and more arousal to their particular voice.

NC: So higher for their species and still higher for themselves?

RT: Exactly.

NC: Is there any kin effect?

RT: That's a good question, and I don't know the answer. In general, kin relations in birds are poorly developed—they often don't even nest next to their relatives.

But in principle I thought you could run a Gur and Sackeim experiment on birds, where pecking could substitute for pushing the button on the computer. You would train them in a reward system to peck when they recognize their own song.

NC: So how do you get to self-deceit from this?

RT: Well, you would manipulate them once again by, for example, subjecting some birds to negative experiences like losing fights, which you could rig by matching them with animals that are somewhat larger than them. And similarly, others would get to win fights. And then you could see if there's a tendency to deny self.

NC: You might be interested in a book that's coming out by a very smart guy, James Peck, a Sinologist, called *Washington's China*, in which he does a very in-depth analysis of the national security culture. It's about the imagery of China that was constructed in Washington. He went through the National Security Council literature, background literature, and so on, and he does both an analysis of content and a psychological analysis. I was reminded of it the whole time you were talking.

What he says is that there are elaborate techniques of self-deception to try to build a framework in which we can justify things like, say, invading or overthrowing the government of Guatemala, on the basis of some new objective. And it's done by making everything simple. You have to make it clearer than the truth.

RT: Right.

NC: And as this picture gets created internally and built up by each group of National Security staffers, it becomes like a real fundamentalist religion, showing extraordinary self-deceit. And then you end up with the Cheneys and the Rumsfelds.

RT: I've been appalled lately when I pass a newsstand and there's some article, "China, the Next Threat," saying, "Now we've got to mobilize all our energy against China"—and they're talking military.

NC: That's interesting, because the threat of China is not military.

RT: Exactly.

NC: The threat of China is they can't be intimidated—in fact it's very similar to what you've described. Europe you can intimidate. When the United States tries to get people to stop investing in Iran, European companies pull out, China disregards it.

RT: Right.

NC: You look at history and understand why—China's been around for four thousand years and just doesn't give a damn. So the West screams, and they just go ahead and take over a big piece of Saudi or Iranian oil. You can't intimidate them—it's driving people in Washington berserk.

But, you know, of all the major powers, they've been the least aggressive militarily.

RT: No, the obvious threat—I mean, the obvious "threat"—is economic.

NC: And I think they plan it carefully. Like when President Hu Jintao was in Washington. When he left, he was going on a world trip. The next stop was Saudi Arabia. And that's a slap in the face to the United States. It's just saying, "We don't care what you say."

RT: Right.

NC: I'm sure it was planned. That's the kind of thing that intimidates. It's a little bit like a gorilla pounding its chest.

RT: Yeah, exactly. More power to them. ∞

Eight

On Dreams
Michel Gondry and Robert Stickgold

The filmmaker and the Harvard psychologist explore memory, creativity, and the science of sleep.

When Michel Gondry was an artist-in-residence at MIT in 2005, a graduate student gave him copies of the readings from one of Robert Stickgold's sleep and dreaming classes at Harvard. Gondry was intrigued by what he read and how it might inform his newest film project, *The Science of Sleep*. The two didn't meet until spring of 2006, when Stickgold was invited to a screening. They hit it off at the dinner that followed and continued their conversation—from Freud to Fellini—at the Algonquin Hotel in New York City.

Writer and director Michel Gondry won a 2005 Academy Award for the original screenplay *Eternal Sunshine of the Spotless Mind*. Lauded for his inventive visual style, Gondry started out making music videos and his 1994 Levi's "Drugstore" commercial was listed in the Guinness Book of World Records as the most award-winning commercial of all time. Gondry's 2006 film, *The Science of Sleep*, premiered at the 2006 Sundance Film Festival and chronicles one man's inability to discern dreams from reality.

Robert Stickgold is associate professor of psychiatry at Harvard Medical School and works in the Laboratory of Neurophysiology at the Massachusetts Mental Health Center, where he studies the role of sleep and dreaming in memory consolidation and learning. In 2000, Stickgold presented a unique research method using Tetris, which demonstrated an ability to predict peoples' dreams. Stickgold is the author of two science-fiction novels and many popular articles on the science of sleep.

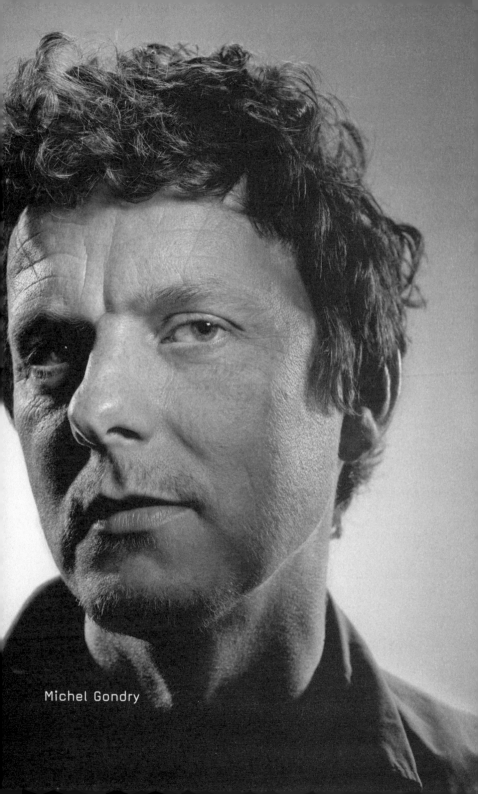
Michel Gondry

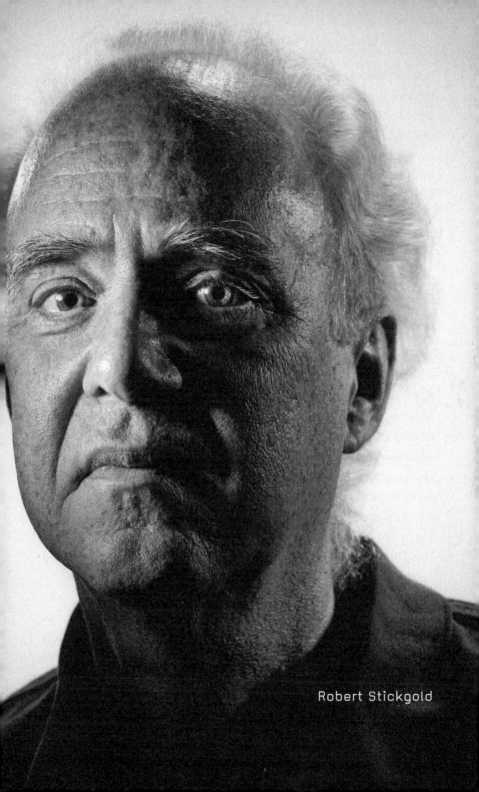
Robert Stickgold

Robert Stickgold: I have this problem. If I go to a party and introduce myself to someone and they ask me what I do I can either say, "I'm a sleep and memory researcher" or "I'm a sleep and dream researcher." If I say I'm a sleep and dream researcher, the reply is guaranteed: "Oh, I had the strangest dream last night."

Michel Gondry: Right.

RS: Then they tell me this absolutely banal dream. "I dreamed that I was at work and I couldn't get the key to open my office and then suddenly I was at home eating dinner—what do you think that means?" My stock answer is "You are one sick cookie." And they always get upset when I say that, as if I know something. It's funny, this desire to interpret dreams.

MG: And there are so many books devoted to that. I think it's complete rubbish, the way they interpret the symbolism in dreams. It's silly. But people need to be reassured, I guess. People need to be told who they are.

RS: But it's interesting that they want to use their dreams to do this, because as a scientist I can tell you that our meta understanding of dreaming as a process is phenomenally primitive and uncertain. Dreams themselves have a huge element of uncertainty and lack of clarity, and our understanding of them when we wake up is often equally uncertain and unclear. So there is this sort of recurring uncertainty from within the dream, to looking at one dream, to looking at dreams in general.

MG: Did I tell you when we met [about] this way I have of trying to relate my dreams to my memories?

I take a piece of paper and I write down the basic plot of the dream, and then I write what memories the dream made me think of. So, for instance, I remembered I was in this room and there is this weird bench—it was a bench that should be outside. And then my son appeared for no reason. So I wrote "the bench" and then I remembered that this bench was outside the coffee shop I walked by the previous day. In this coffee shop I remember having my second date with my girlfriend, when I told her that I was a father. So when I look at that my son and the bench get connected.

RS: So when you're asleep, your brain has access to certain sets of memories, and it takes those memories and somehow combines them and takes elements from them and constructs this dream. And then you wake up in the morning, you recall the dream, and you try to untangle it back to those memories. But your brain is in a very different state when you wake up in the morning than when you're having the dream. So the question, from the scientific point of view, is: Is the narrative that you're constructing after you wake up the same narrative that went into creating the dream?

MG: Yeah, I guess it is different, and maybe the difference comes from the way you put it back together, because that's a creative exercise, and that process can say a lot about you.

RS: Right—discovering or assembling a meaning from within the dream.

MG: Yeah, it's like found footage. You immediately want to try to decipher meaning from it, and formulate a narrative—our urge toward having a beginning, middle, and end is extremely powerful. But then you also want to know who was the filmmaker? And how is he or she related to these images?

RS: Well, it's interesting—we've tried to look at three hundred dreams from thirty different subjects and had them do this exact exercise, saying, "Okay, what is the thing that happened in waking life that you think caused this to happen in your dream?"

MG: Mm-hmm.

RS: And, like you, they have very strong confidence that there are strong relationships and "oh, this bench was in the dream because of this." But then what you find is that they don't have the whole story of that memory. They have one piece out of it.

MG: So they do a selection that's subjective.

RS: Something does a selection, and what we find is that the emotion and the theme are the strongest connections, and then maybe some objects—but people and places are very rare. So it's as if the brain is taking a sort of slice. Like when you're filming, it would be as if you came in with another camera at a ninety-degree angle that showed you a completely different way of seeing the scene than the way you'd been watching it from the front and all of a sudden you realize this person is way behind that person and it looked like they were closer together than they really are. And so you get a different slice when you dream, and it might be that there is a very clean way it fits together when you wake up that is actually very different from how it was put together in the dream. I don't know how to sort that out. I mean, as a scientist, that's the frustration.

MG: Sometimes I analyze the dream as I'm dreaming because I'm lucid-dreaming; I'm aware. And sometimes I dream in a loop where I'll come back to the same spot and redo exactly the same thing, repeating the same story again and again, four, five, six times. So I have the sense that things can occur simultaneously in sleep, and then only when you recall them do you create a linear narrative for them.

RS: Right, yes. But you know, it's very funny that you talk about how you do the same thing five times over again in a dream and that that's not the way it works in waking life. But that's what you do in your job.

MG: Mm-hmm, yes.

RS: You shoot the scene, you go back and you shoot it again, and you shoot it again. And so I want to say, "Aha—I know why you have those dreams!" But I don't know if that's the reason or not. It's a very frustrating thing that it's much easier to find an explanation that sounds and feels good than to know it's the right one.

MG: Maybe you're right though. It's a sickness—when you are a filmmaker, you don't feel you can enjoy life if you don't record it.

RS: But that's a very different life, the one that's recorded.

MG: Yeah. But the trouble is to try to make it well but to keep some life when you record it. For me, the best way is to de-prepare for the part. People prepare themselves; they have an idea of how things should be, and it's often very conventional. I need to erase that to get who they are. So to achieve that I change the camera, I change the direction, I change something at the last minute. Nobody is prepared, and everybody is acting on their toes, and they are in a very creative process because they have to readjust.

RS: So you have just described the dreaming process. That is exactly what dreaming is. Dreaming is breaking you out of your normal set, and it does it exactly the way you do it, by deprogramming. It shuts off your access to memories of events in your life.

MG: Mm-hmm.

RS: All those events that make up your life, that's your narrative.

MG: So you don't have access to those in your dreams?

RS: You don't. You have access to pieces of them.

MG: So, then, you're not the same person when you dream—because it's memories that form people's personalities.

RS: You're the same person looked at a different way. So when you tell the story of your life to someone, it has a strong temporal flavor. You talk about what happened over time: "I did this and then I did that."

MG: Yeah.

RS: In your dreams that temporal element of memory seems to be completely disconnected.

So you can't say, "First I did this, and then I did that." What you say is "I did this, and that was sort of like this other thing," and you move through a network that is controlled not by time, but by the fact that there was a bench, and on this bench you had a conversation about your son.

MG: Right.

RS: Something happened on that bench that has to do with your son; here is something else important about your son, and so on— and it can come from anywhere in time. When you tell the story about this event, about sitting on the bench talking to this woman, it's your second date and you finally tell her you have a son.

MG: Yeah.

RS: And you know this story well. You've told it a hundred times in your mind and so it tends to come out in a classic, standard way. But when you dream, that narrative, which we would refer to as an episodic memory—a memory of an episode in your life—is completely dissociated, all the pieces are broken apart. And this is a classic phenomenon in dreams. By and large, people just go right along with the flow of it, and those sorts of discontinuities are not problematic. It's no more of a problem than when you are watching a movie, and someone is talking on the phone in Paris, and the next minute he's on an airplane flying over Africa. It's a cut.

MG: Yeah, actually I have been talking about this. People ask me, "Why did you do a movie about dreams?" First I say, "Why not?" And then I say, you don't live with cuts, but you dream with them, so there are these kinds of similarities.

RS: It was Sergei Eisenstein who first did that, who actually showed them in a shootout and then just cut to them falling. Until then it was assumed you had to follow the actors like you do on the stage.

MG: Yeah, yeah.

RS: But it worked from the very first time it was ever done, because we're all familiar with it.

I don't know how much of that style is in fact successful *because* we know it in our dream lives, though.

MG: Mm-hmm.

RS: But it does lend a dreamy flavor to film, so that the movies shift into a more symbolic mode and then give you permission to make that extrapolation back out of the narrative.

MG: It's also about giving the viewer enough space for interpretation. Putting it together with enough room for interpretation that people feel they can come up with their own explanation.

RS: And so the question for a filmmaker is "How much do you want to steer people toward one interpretation?"

MG: Right, exactly. I like to give audiences a guide through the casting and the dialogue but in the end, I leave it open to their personal take on the film.

RS: With many movies, especially Hollywood movies, it feels like there is one interpretation everybody is going to come very close to. And in that respect, they're boring. They might be entertaining, but there's sort of nothing to learn about yourself from it, because you learn about yourself by how you interpret things. Whereas—I remember when you were taking questions from the audience after the screening, and someone asked you the question "What was that supposed to mean?" I cringe for you when you get a question like this, because the answer that you gave was "I put that out there for you to find a meaning of your own" and there are always people who are profoundly uncomfortable with that.

MG: Yes.

RS: Those are the same people who buy the books to tell them what their dreams mean. Everything has to be laid out for them because they are afraid to live with the ambiguity of the world.

MG: It seems to me that the scientific approach is that you don't work with coincidence. If things seem to fit and look like they are connected, but there isn't proof, it's your job to separate them until there is proof. It feels like when I take the dream and I find really very precise and crisp coincidence it makes me think, "Okay, I don't think there would be any other way." Your job as a scientist is to say, "Well, until it is proven, it could be this or another incidence or another connection."

RS: Well, science can be a clumsy sort of process in the end. I mean, it's true, we never assume that correlation is causation. That's one of those things they drill into us from day one: don't assume it's causation. And for some scientists, I think this is absolutely self-destructive, because they go to the extent of sort of denying their own creativity.

MG: Uh-huh.

RS: Because really, science moves forward in meaningful ways because people see correlations that they believe in.

MG: Yeah, they have a vision, and then they demonstrate it.

RS: And then if you're a good scientist, you get terrified, because part of you knows it's only a correlation and there might be no causal link at all. And so I will tell people when I talk about my work: "I know this is a beautiful story. I know it all fits together perfectly. I just don't know if it's true."

MG: Right.

RS: And the trick is to maintain some kind of belief and almost religious fervor about your model while still another part of your brain is saying, "Yeah, well that's nice, but show me the data."

RS: I think the use of dreams as a tool within film and literature is a very tricky business, because usually they put in a dream that's deeply revelatory to the person who has it: they wake up and their world is explained or the dream explains everything, and it's always kind of hackneyed in that way. It sort of feels like a cheap shot; it's a bit of melodrama. You can give a person any dream you want and—

MG: It will help the story.

RS: It helps the story and it sort of feels like the cavalry comes rushing in at the last minute. But you did dreaming in your movie very differently from that.

MG: I think I have been interested in my dreams since I was a very young child. I had a lot of nightmares and so I was drawn to try to look at them, not understand them, but look at them. I didn't want

Michel Gondry and Robert Stickgold

to use them just to make the story move forward. It's more like a journal—they are just exposed, the way I remember them, because two-thirds of them are real, and I put each dream in the context of the event I experienced at the same time I had it.

RS: So you're chronicling dreams. You aren't using them.

MG: Right.

RS: You're describing them as they actually occurred—as you remember them.

MG: Yeah, exactly.

RS: Very different from Fellini, for example, who made an art form out of dreams in his own strange way, but there is no suggestion that he was being true to dreaming.

MG: Oh, but I wouldn't make this judgment on his work. When I see a David Lynch movie, for instance, I'm not sure he's actually had this dream or not, but it *feels* truly dreamlike. They don't feel like they move between awakeness and dreaming, they *feel* like a dream. I remember watching *Lost Highway* and literally "waking up" at the end of the film. There is something geometrical about the way this movie was constructed—like the two characters switching in the middle of the story for no reason—that was really the feeling of a dream.

RS: So that's another way to do it—make a whole movie as if it were a dream.

MG: A lot of the time I'm directing my dreams, and it's totally clear I'm trying to make things happen. For example, I'm thinking, "If I open this door I'm probably going to get scared and I want to feel that and see how it is." I remember being in a dream and trying to internalize the definition of the image. I don't always realize that I'm dreaming, but often I do, and I remember clearly looking at the image and saying, "It is not as sharp as thirty-five millimeter; I would compare it to sixteen millimeter, although it's getting very blurry on the side." Or I'm in my dream and I'll say, "Okay, let's explore more," and I go into a recording studio and listen to the sound of the treble and think, "Ah, this is interesting. I can really dig into the sound to get more substance in it."

RS: Do you always sleep on your ideas before you film them?

MG: Oh, yeah. Well, most of the time it's just because I shoot over many days, so I have many nights of sleep.

RS: But do you feel that that changes things?

MG: Yeah, sometimes. Sometimes I can find a solution. It's funny, actually. I tend to find more solutions by dreaming than by working on it before I go to bed. For instance, dreaming an idea allows me to use it without questioning it—as if the dreaming process were legitimizing the idea.

RS: So you think those are useful dreams?

MG: Yes, those are useful. At least to the dreamer, but otherwise . . . I don't know—maybe the dream is just like an appendix?

RS: Well, the most optimistic of estimates says that people can remember 2 or 3 percent of their dreams. So 95 or 98 percent of your dreams you never recall when you're awake, and yet you can ask, are they useful even though you don't remember them? My view is that they almost undoubtedly are, because dreams are in some way a shadow of processing of emotions and memories that occur while we're asleep. And just like there are all sorts of things that happen during the day that we don't remember that affect us, all these dreams are, in their own ways, tuning, if you will, our memory circuits and our associations.

MG: When you think of the use of dreams, are you also trying to put them into the context of evolution theory to find out how it benefits a species and so on?

RS: You can do it two different ways. You can talk about dreaming as an adaptation, something that evolved and was selected because it increases survival, but you can also talk about uses that, in the terminology of evolutionary biology, would be called a spandrel. For

Michel Gondry and Robert Stickgold

example, the sound of your heartbeat didn't evolve because it was useful. It's just that a heart evolved that pumped blood and when it contracts it just happens to make a sound. That sound is very useful to us now—your doctor takes the stethoscope and puts it onto your chest and listens to your heart and gets a lot of information. So in that respect, the heartbeat is very useful and might even help you survive, but that's not why it's there.

MG: But on the other hand, we have a system to stop the sound of the blood, otherwise we would hear it in our heads constantly.

RS: Right, the sound of your blood in your ears would start to wipe out all other sounds. But what that says is that we can talk about dreaming in those two different ways. We can ask, "Did dreaming evolve because it confers a powerful mechanism for processing memories and emotions that makes us more functional and better able to survive?" Or "Is dreaming this goofy byproduct of something else that is actually useful?" I mean, look at tarot cards. They lay out these cards and they tell you what it means. Now you and I agree on one level that that is all nonsense, right?

MG: Yeah.

RS: But that doesn't mean they aren't useful. That doesn't mean that people can't, by constructing a narrative around those cards, explore their beliefs, their emotions, their plans. When there are all these sorts of loose descriptors, your mind goes into a more hyperassociative space. And what the tarot card reader does, what the daily horoscope does, is just push your mind in one direction or another.

So we can come back to dreams now and say, are they just there or are they useful? I think there are ways to use dreams that are very destructive. I think some of the old, not so much the modern, but the older classical Freudian interpretive structures were highly destructive. You have an erotic dream and it's a sign of real pathology. You have almost any dream and it is a sign of real pathology.

MG: Yeah, I have a big issue with that.

RS: So it's not a question of whether dreams are inherently useful or destructive. Like the media, they can be used in different ways, and they have that power—and that is their value *and* the danger, right? It's the same thing with making movies. Movies can help people understand important things. They can be demagogic and pervert large segments of the population, and they can be totally agnostic and leave people wide open to decide what they want—the medium doesn't define which of those happens. The medium just provides the potential for any of those.

MG: Mm-hmm.

RS: So it's not about the dreams, it's about the interpretation process that puts that value onto it.

MG: Right, exactly. I'll probably dream about that tonight. ∞

Nine

The Truth of Fiction
Janna Levin and Jonathan Lethem
The cosmologist and the novelist meet up to talk about reality.

When Janna Levin began writing *A Madman Dreams of Turing Machines* it was a work of nonfiction. But she realized, as her subjects Gödel and Turing had, that the tools of nonfiction—or those of scientific inquiry—were insufficient for discerning truth. As a novelist, Jonathan Lethem traffics regularly in different degrees of truth and is similarly fascinated with what constitutes reality. In early 2007, the two met for lunch at the National Arts Club in New York to talk about this elusive concept—its guises, its enchantments, and how we know it when we see it.

Janna Levin is a theoretical cosmologist and a writer. Her research involves the origin of the universe, the shape of space-time, chaos, and black holes. She is the author of two books, most recently the novel, *A Madman Dreams of Turing Machines*. Levin has worked at the University of California, Berkeley, and the University of Cambridge, and has served as a scientist in residence at the Ruskin School of Drawing and Fine Art in Oxford. She is currently a professor at Barnard College of Columbia University.

Jonathan Lethem is the author of seven novels, one novella, two short-story collections, and a volume of essays. Known for defying traditional genre conventions, Lethem instead mingles them to achieve a distinct and original voice in American fiction. His novel *Motherless Brooklyn* won the National Book Critics Circle Award, and in 2005 he received a MacArthur "Genius" Award. His work has been translated into nearly thirty languages. Lethem's newest novel is *Chronic City*, set in Manhattan. He lives in Brooklyn and Maine.

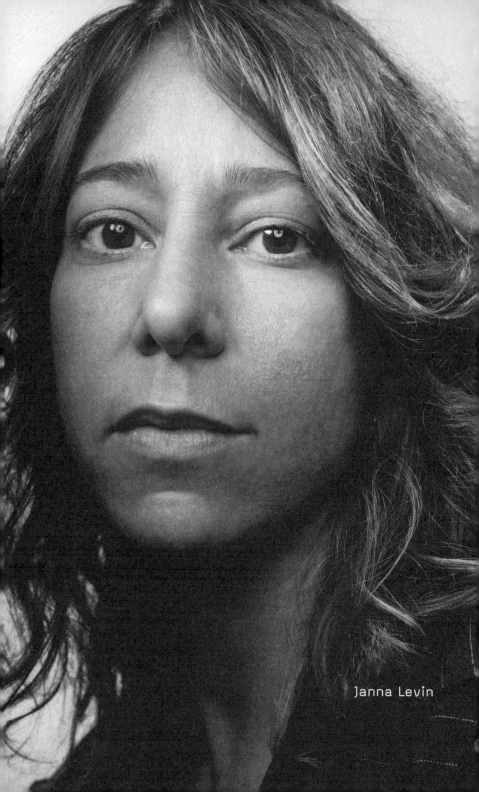

Janna Levin

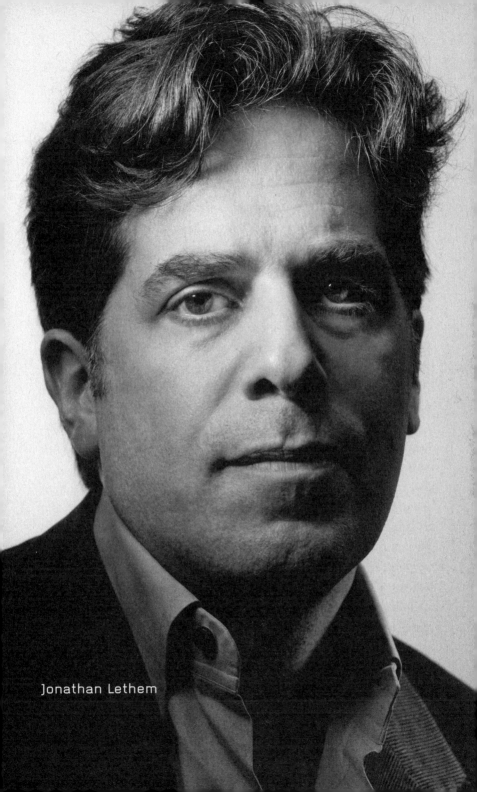

Jonathan Lethem

Janna Levin: I've found it very interesting that in all of your novels, there's something fanciful. There is always this element that's not real, even though there's a very realistic quality to your writing.

Jonathan Lethem: Well, I think one place that comes from is that my father was a painter, and I was trained as a visual artist.

Levin: And that shows up in *The Fortress of Solitude*.

Lethem: Yeah. The main character in *Fortress* has a painter-father, too. And we both took our father's work and incorporated it into a worldview. In the case of my father's painting, he always combined representation with the imaginary or the fantastical. And I took this as a kind of basic condition of art that, in a way, was inherent in my worldview before I could ever have questioned it. Art consisted of a combination of observed elements and—the other word is always harder—the imaginary or fantastical or metaphorical. In some ways I think "metaphorical" is the word that captures it best for me, because I've come to see that one thing all of my writing has in common is some element of imagery that moves out of metaphor and into the real space of the story. Whereas in someone else's story, the people might feel that they're superheros, in my book the characters actually get to try out being superheros. It's a kind of literalness about metaphor.

Levin: That's an interesting way of putting it—a literalness about metaphor. What you said about the confluence of the observed and imagined is very interesting because that's of course what every novelist must do, at some level. We're disappointed if novelists don't combine the observed and the imaginary.

Lethem: I agree completely. And I always find it remarkable that people are praised for their realism in books that sort of sublimate the imaginary or metaphorical elements because it seems to me that if anyone were actually ever handed utter realism, which is to say a kind of a transcripted human conversation—

Levin: Which our readers are going to get!

Lethem: —they'd be bitterly disappointed. And yet this notion that that author ought to deliver reality in some kind of document persists as praise. But I think it's inherent to art, and certainly to the art of fiction, that invented and observed material go together. And what I do, I guess, is kind of make the point of collision rough and obvious instead of smooth or sublimated. I have a tendency to want to make it unmistakable and, in some cases, kind of uncomfortable for the reader. So, as you say, *Fortress of Solitude* invites you to feel that you're in one kind of depiction of reality.

Levin: And then suddenly there's something—

Lethem: An eruption of magical possibilities.

Levin: Right. But what's so disappointing about reality? Because I know what you're saying, but what is it that we need to be repackaged and rephrased for us so it hits our pleasure centers better?

Lethem: I like your devil's advocacy, but I don't think that's the problem. The trick, I think, is that the external material surface of reality, the documentary depictions that I'm saying would be inadequate, are only one version of reality. That, in fact, consciousness and the emotional, the social, the—to take a term from your work—*relativistic* aspects of experience are much, much more magical and dimensional and inexpressible than the raw tools of documentary realism could ever capture. So what we experience moving through life is a kind of collision in our consciousness of the imaginary and the documentary.

Levin: Yeah, that's interesting. Of course, in my scientific work, there's only one reality. There can't be ambiguity in my scientific research—it's all about the reproducible answer. I want the calculation that is going to precipitate an answer that's definitive, that both you and I will agree upon regardless of whether we're from different countries or we're educated differently. Yet it's exactly the aspect of experience that's so important, even in science. And that has to do with all those things coming together in the way that you described—how you perceive what you're doing, why you feel it's important. What does it mean for your life or other people's lives? Why do I think this is worth funding and researching and then relaying to the world? And those questions cannot be answered by a series of facts. I don't do straightforward nonfiction in my books because I think it's slightly dishonest. It's never straightforward nonfiction; there's no such thing.

Lethem: It hides the fact that it's conjured.

Levin: Right. Unless I write down only the core mathematics and nothing else, and even then my approach could be really unique and therefore impact differently, and so on. But once it comes out in words and in language, it's something totally other. And so this was a real motivating impulse for me with *A Madman Dreams of Turing Machines*. I wouldn't call it fiction exactly, but I certainly wouldn't call it nonfiction. It hovers in that divide.

Lethem: Well, one of the underrated aspects of novels per se, one of the forms of pleasure that we readers derive from reading fiction that is least discussed in traditional literary criticism, is factual material. People thrive on finding great chunks of information on how the world works in their fiction. One of the great secrets to the crime drama is that readers are almost always inadvertently thrilling to descriptions of how, for instance, a bank operates. These are the sorts of things that ordinary novelists feel that they're not allowed to talk about or get interested in—they're supposed to be concerned with the emotional or psychological lives of their characters and would never stop to tell you at what hour the teller counts her drawer and moves it to the back of the bank. And yet we're all hungry for those pieces of information about our world. We're nourished, without even noticing it, by this genre that's devoted to telling us quite a lot about them.

Levin: Right.

Lethem: And, of course, part of this is a fiction writer's thrill to genuine jargons as well. And by genuine, I mean necessary jargons. The language is a specialized reality, and there are these incredibly specific differences in the way things are described from the inside and the outside. And those have such power of persuasion. It's incredible how, for instance, a policeman says of another policeman, "He's a good police," which a citizen would never say. Just as I noticed when I began gathering material for *As She Climbed Across the Table* that physicists would say of another physicist, "He does good physics." I was so turned on by the specificity of that. It seemed to imply such an inside to me, and I wanted to move within it and experience it. So I began doing the best I could.

Levin: And as you'd mentioned to me before, you're interested in the response to things. You're not necessarily interested in whether or not you can really make a false vacuum universe in the laboratory, for example. As a scientist, I am interested in whether or not you can really make a false vacuum universe in the laboratory. You either can or you can't. It's either true or false. Yet as I was writing *A Madman Dreams of Turing Machines* I was forced to contest this simplistic faith in truth. I mean, here I was so drawn to these mathematicians who prove that there are some truths that can never be proven to be true. The idea that even mathematical truth is elusive is very unnerving for somebody like me. I'm a realist. I believe that you're not a projection of my imagination. I believe there is something real about you that I'm perceiving. The physicists and mathematicians whom I describe in this book weren't so sure that you couldn't be reduced to a projection of their imaginations. They had no philosophical or scientific argument to prove otherwise. And if science couldn't prove definitively that you were sitting there, then they couldn't believe that you were sitting there. Beyond that philosophical stumble, Gödel and Turing proved that even mathematics was incomplete, that arithmetic could not capture every true fact about numbers. What's curious is that in my personal life, or in my life in relation to art or novels, I have no problem admitting that truth is a muddled, imperfect notion. I well appreciate that we can get closer to the core of an idea through metaphor or imagery. That, as you say, there is a collision in our minds of the factual and imagined that creates a reality of sorts. But science can be so gratifying precisely because scientific truths are not ambiguous. I don't like using science as metaphor because I think it betrays a certain aspect of what's precious about scientific inquiry—precisely that lack of ambiguity. So here I was with this rub of making truth metaphorical, making truth elusive, seeing how far it can go. And I wanted the book to hinge on that conflict, to celebrate the obsessive pursuit of mathematical truth even as it argues no truth can ever be captured in its entirety.

Lethem: What's interesting to me is that you're so driven to pursue the kinds of truths that remind me of those the artist inherits automatically, which are paradoxical truths. The ones that can only be surrounded, that can't simply be described. And both of your books include attempts to surround paradoxical truths with enough understanding or visualization or description that they can be apprehended because they can't be looked at directly. In a way, that's my whole job. And so it isn't that I'm uninterested in truth, it's that I never think I'm going to get there by compiling other truths. I think I'm going to get there by compiling responses and paradoxes and evocations.

Levin: So you are getting to some level of truth.

Lethem: Of course.

Levin: You wouldn't say truth is totally relative.

Lethem: No, I wouldn't. My instinct about the scientist who decided that it was possible that we were projections is to say he's made the purist misunderstanding. For me, the truth is always about this muddying of actuality and metaphor. We live on a mingled plane. Certainly, my characters dwell there, and I, personally, dwell there. So the false obsession with purity is what has to be chased away.

Janna Levin and Jonathan Lethem

People fancy that, say, Raymond Carver was a pure realist—well, this is nonsense. The very tools he used were corrupted, dripping with metaphor, because language itself is big chunks of metaphor that we're moving around and operating with. And I think it's the glorious impurity that has to be reasserted.

Levin: Yeah, the fabric of language is symbolism. That's all it is. It's verbals, it's a sound to symbolize a certain thing.

Lethem: Exactly.

Levin: You said something very interesting, that you could be so purely tied to an attempt to get toward truth or an attempt to adhere to logic that you're totally lost. I find that irony beautiful, that the people whom I wrote about, who stuck the closest to logic, were the furthest afield, the most lost. They became utterly confused and ended up drawing conclusions that even they knew were wrong. Logic took them away from truth. If I really believe that all I know is something that's scientifically verified—I don't really know that I'm touching this table, all I know is that I'm experiencing what I've learned to call pressure against what I think is a hard surface—then I just get further and further away. And I think that's so remarkable. We need the fuzziness of imperfect thinking to function.

Lethem: Exactly—there's an enormous amount we can agree on about our experience in order that we can dwell in a meaningful place to pursue the elusive truths, accepting that there's a good enough description of a table that we can have this meal.

Levin: Right. And that's exactly why Gödel gets lost. He gets totally lost, he's paranoid, he's schizophrenic, and yet he isn't really insane. I mean, you can see the way he lives his life as a consequence of being so logical. It's not that he believes that there are pink flamingos flying in his apartment. Rather, he's so tightly sticking to logic that he becomes very paranoid and comes to believe things that the rest of us don't believe because we're so illogical and so irrational.

Lethem: Right.

Levin: And so even his bizarre and dire suicide, this attempt to starve himself to death, was, I began to feel, actually this logical, inevitable consequence of sticking to the rules.

Lethem: Right. He refused to take consolation in assumptions.

Levin: Absolutely. And, of course, the ideas of Turing and Gödel became central to the invention of the computer and, ultimately, the ambition for artificial intelligence. Yet one thing that their theorems require is that artificial intelligence programs are not perfectly logical, because then they can't do the kinds of strange things that we do. It's actually embedded in their theorems. That's a remarkable consequence. And still it took the computer-science community a very long time to say, okay, we can't literally program these things. We have to watch them somehow evolve in all their complexity.

Lethem: Right. And, of course, as a reader of science fiction as a teenager, I saw Turing's name thrown around all the time.

Levin: See, I don't read science fiction.

Lethem: The Turing Test became a kind of cliché in science fiction, which loves to dwell on the difference between man and machine and possibility of perfection. So he'd offered something kind of irresistible.

Levin: Like robots don't dream—or they do dream—of electric sheep?

Lethem: Androids dream of electric sheep.

Levin: Oh, right—ha! I knew I'd get it wrong.

Lethem: One notion that I was very interested in—and maybe I was reading your books and finding these moments because it's a preoccupation of mine at the moment—is your awareness and sensitivity to questions of originality and collaboration and individual achievement, which are, of course, obsessions of the scientist maybe even more than the artist. But certainly they're mutual obsessions. And you had Gödel talking about how someone else would have thought of it if he hadn't, and it all broaches the question of whether the ego of the individual, artist or scientist, is essential, and whether the work is truly a collective enterprise.

Levin: Well, I think in science it's a very interesting question. If you really believe in objective reality, then you don't matter at all. If it hadn't been Einstein, eventually it would have been somebody else. And if it hadn't have been Gödel, eventually it would have been somebody else. And so scientists are playing this difficult game with themselves, with their own egos. They want to be the most brilliant, get there first, be accomplished, and yet at the end of the day, they have to say to themselves, "But it really didn't matter that it was me, and my marks cannot be left on this in any way." It's even in the way scientists write—all scientists begin their papers in exactly the same way. I mean, it drives me insane, but there's a certain—I'm not going to say charm, to it, but I understand it.

Lethem: There's a beauty to the sort of fierce belief in pure thought.

Levin: That's right. Even if the paper has a single author, "*We* show that . . ." Yet the ego tensions in the sciences are outrageous. I mean it's totally outrageous. And so, from a storytelling perspective, it's interesting because there's conflict in that experience. There's conflict in discovering. And conflict usually—well, do you think conflict drives our great stories?

Lethem: Well, I suppose it does. And I suppose there's tension between collective or communal reality and the starkness of individual experience, and also the possibility of individual, transportive revelation in the artistic epiphany.

Levin: See, I don't believe artistic experiences would be reproduced by somebody else.

Lethem: Right, well, Saul Bellow's novels really wouldn't exist if he hadn't written them, in the same way that monkeys couldn't sit down and write Shakespeare. Well, in fact, humans couldn't sit down and write Shakespeare; Shakespeare had to exist. And Saul Bellow or Jonathan Lethem, with their immensely particular, eccentric form of biases, had to exist. I mean my own irritability, my own distractibility, are in my books in a way that's just absolutely present. And yet at the same time, it's the things that become fetishized as innovations, the avant-garde gestures and so-called experimental breakthroughs in the arts, which are the aspect that is more analogous to scientific discovery—

Levin: And which would have come anyway.

Lethem: They would have always come anyway.

Levin: Right. Realism would have given way to abstract expressionism.

Lethem: And there are always precursors for everything, almost to a comical degree. If you see someone being praised, say, in the Sunday book supplements for inventing something in fiction, be guaranteed that there are a hundred people who did that before. It's just in the nature of the ostensible experiment or avant-garde. But then again, the brush stroke or the syntax and so on is individual. It's almost as though the confusion is over intellectual property, and the protections that are offered to art misunderstand its essence because it's so much more intimate and so much less quantifiable or commodifiable.

Levin: Uh-huh. Well, you have to, I think, at some stage make an admission about who you are and what you bring to the table.

Lethem: Right.

Levin: Something I find particularly interesting is that science, I think, is the last realm in which people talk to each other seriously, with a straight face, about beauty. Visual artists would never say that's a beautiful piece of work, not in really contemporary, cutting-edge art.

Lethem: That's a very difficult phrase. After modernism, beauty is terrifically suspect.

Levin: Right, absolutely. And it's considered kind of provincial to aim for something beautiful. We're not doing pretty pictures here; we're doing something else. But in science, we really hold on to beauty and elegance as the goal because, for reasons that I think nobody fully understands, it's a good criterion for distinguishing what's right from what's wrong. And if something is beautiful and elegant, it's probably right. Occasionally, you'll see something that's so beautiful and so elegant, and it's not right, and you can't believe it's not right.

Lethem: Yeah.

Levin: And I don't understand that. I don't think anyone understands that. You can make arguments that maybe it's only human perception, that we impose patterns on a vast sea of complicated things.

Lethem: Right. Not so totally different from the insight that one of the things that people were selecting for without naming it when they thought a person was beautiful was—

Levin: Symmetry.

Lethem: Symmetry. And now we take that so for granted. But in fact, this was not something anyone really recognized. They just sought to describe an evoked response.

Levin: Right, but of course our responses evolved out of a series of complex steps, starting with simple laws of physics. And it's a marvelous thing that we're even able to know certain things that we inherit but are so far away from. What was Einstein's quote? Something like, the most miraculous thing about the universe is the fact that we can understand it. And it really is miraculous. We're a pretty humble species; we've been around for a very short time, yet we know things about the origin of the universe 14 billion years ago. And it doesn't have to be that way. Though again, you can make the counterargument that yes, it does have to be that way because we evolved from a series of steps that started with the origin of the universe. And so somehow, all of those physical and mathematical processes became imprinted on our logical networks. According to somebody like Turing, we really are just biological machines, and we're coded and programmed by the laws of physics as surely as if somebody had sat down and written that code. So maybe the fact that we are putting the name beautiful to it, and elegant to it, is, as you said, just because we're identifying something that's been selected for in an evolutionary way.

Lethem: Right. It's a way of connecting A to B, of realizing a truth. And when I talk about pursuing epiphanies, in a sense it's another word for truth. When I made that distinction and said, well, it isn't that I'm not interested in truth, it's that response or epiphany is the building block toward the truth or paradox evocation, those are the ways I chase my truth.

Levin: Something I also thought about is the act of guiding a reader toward a truth you want them to recognize. I'm getting more and more interested in film right now, specifically these past few months. And I think it has to do with thinking about not just other ways of writing, but other representations of reality, or other representations of the truth or of experience.

Lethem: People take it as a given that the world is presented "as is" on film. When in fact, optically, it's very unlike what our eyes, and our experiences, present us with. You might be interested in reading the essays of Stan Brakhage, a highly experimental filmmaker who tried to start at the beginning again and not take the narrative construction, the editing assumptions, and the camera-placement assumptions of traditional film for granted, but began again at optics and asked how we can make film more like what it's like to look around. His films have this constant movement. They're almost—

Levin: Oh, interesting. Unbearable.

Lethem: —almost unbearable at times, but they're abstract art.

They're like a Kandinsky painting. And in that sense, they seem to derive a connection to—

Levin: Actual experience. But there's that irony again: The closer you try to get to the actual experience, the sort of more abstract and removed it becomes at the same time.

Lethem: Right. Absolutely. ∞

Ten

On Music
Daniel Levitin and David Byrne

The neuroscientist and the singer/songwriter meet up to discuss the sound, the brain, and music.

David Byrne and Daniel Levitin first met in 2006 in Montreal at the Future of Music Policy Summit. While in town, Byrne paid a visit to Levitin's Laboratory for Music Perception, Cognition, and Expertise at McGill University. The two decided to meet again at Byrne's studio in New York a few weeks later, where they continued their discussion of composition in art and music, trance states, and live versus "dead" performances. Soon after, at STK in New York's Meatpacking District, they traded ideas about music, language, and memory. One of many unanticipated and happy outcomes of this Seed Salon was that Byrne invited Levitin to accompany him on electric guitar at a show in New York City.

Daniel Levitin is currently the James McGill professor of behavioral neuroscience and music at McGill University and author of the *New York Times* best sellers *This Is Your Brain on Music* and *The World in Six Songs*. For ten years, Daniel Levitin worked as a session musician, sound and recording engineer, and record producer with numerous bands, including Santana and Blue Öyster Cult. He also served as director of A&R for 415 Records and has been awarded seventeen gold and platinum records over the course of his career. In 1990 Levitin returned to school to pursue his PhD.

David Byrne is best known as the lead singer and songwriter of the seminal band Talking Heads. He has also led an extensive solo career; composed music for dance, theater, and film—receiving an Academy Award for his work on *The Last Emperor* soundtrack—and founded the record label Luaka Bop. In addition to directing videos and feature films, he has published and exhibited his artwork internationally over the past decade. Byrne is also the author of five books, including, most recently, *Bicycle Diaries*.

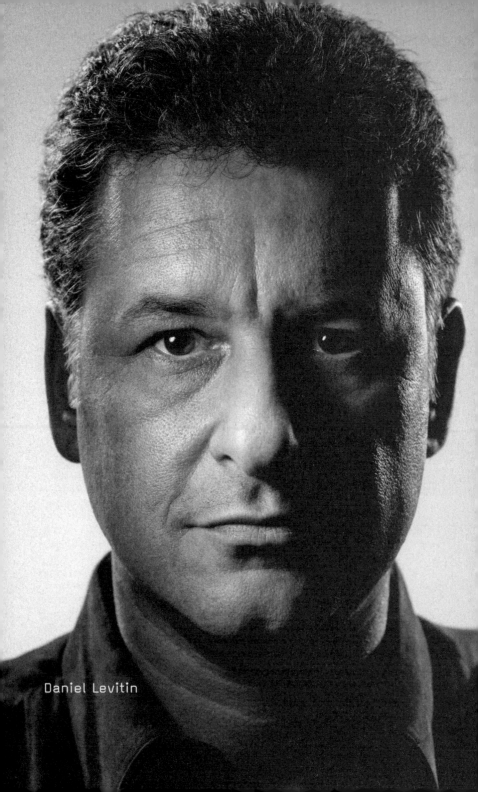
Daniel Levitin

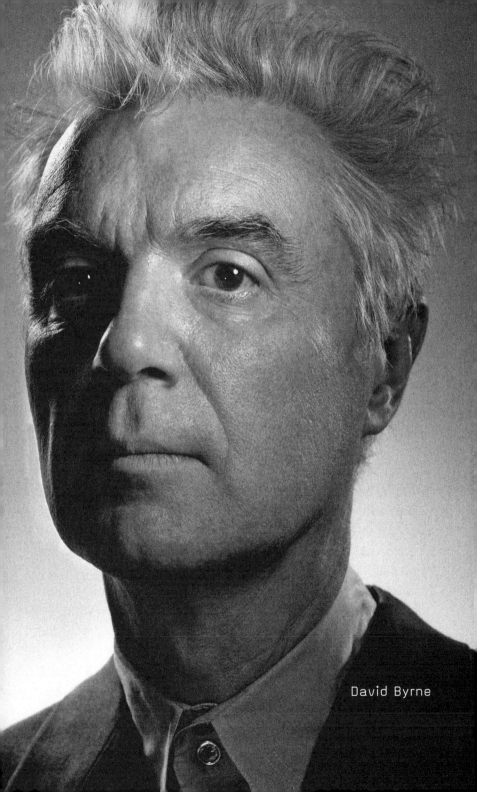

David Byrne

DAVID BYRNE: So, in the penultimate sentence of your book, you write that music is a better tool than language for arousing feelings and emotions.

This ties into what we were discussing a few months ago, about music and visual art bypassing the filters that language seems to get snagged on, in emotionally affecting you.

DANIEL LEVITIN: Yes.

DB: When somebody tells us what this song is about, or what this painting is about, we're kind of stuck because talking about the art, and the art itself, are almost separate areas. The music seems to have straight access to the so-called reptile brain, and we feel it immediately. But often it's also touching all kinds of other parts of the brain. If it has lyrics, there's language in it. If it has a strong rhythmic element it's touching what you would call the motor parts of the brain and muscle. All kinds of stuff is involved. How do you think this all happens?

DL: My guess is it starts with trying to unite rationality with irrationality.

DB: I'll bet you get resistance, too, from people who say you can't analyze this.

DL: Well, I remember a quote from Alan Watts, the philosopher. He wrote a number of books on Eastern philosophy in the seventies. He said that the problem with science is that when it wants to study the river, the scientist will go to the river with a bucket, take a bucket of water out, bring it to the shore, sit there, and study the bucket of water. But of course that's not the river.

And you know a lot of people have tried to study music by getting rid of everything except pitch or everything except rhythm. Or by using very strange, computer-generated sounds, to see what the brain does in response to them.

There's always this tension in science that you want to control your variables and you want to know what it is you're studying. And yet you want to have what we call ecological validity, which is just a fancy way to say it has to be like the real world. There's a tension between these two, and I've erred on the side of having ecological validity in my own experiments because I want to see the real phenomena.

But getting back to what you were saying about why art can get at some of the things language can't.

DB: Yeah. I mean, there's something about music that seems to touch what we would call irrational, emotional parts of ourselves. As somebody who makes music, you know there are kind of tried and true ways of doing that; there are buttons that you can press that will get emotional responses.

DL: Oh sure, the strings from Hitchcock's *Psycho*. I mean, you play that dissonant discordant string sound, and you know the reaction you'll get.

DB: In a musical performance, whether it's recorded or live, people feel the emotion is coming from the performer, and that's what makes it authentic and true and therefore more upstanding and good. Whereas I would say, yeah, okay, a little bit. But music has attributes that you can objectify. This kind of sound, this kind of rhythm, will generate this kind of emotion even if it's done in a half-ass manner.

DL: In "Lilies of the Valley" you're going way high up in your range. Your voice gets a little thin, and you get a crack in it. And it sounds like you're choked up and you're about to cry.

DB: Yes. And that's the intended effect.

DL: So you try to get that.

DB: It's not that I don't feel it.

DL: But you're consciously aware that if you do that it's going to have this result.

DB: Yes. And you could say, well then, it's all just a trick. And it is in a way, but it's also a craft to convey this emotion; this is how you do it.

DL: Yeah.

DB: It doesn't make it less real.

DL: I agree.

DB: This actually makes me think of some stuff I've read recently about empathy and mirror neurons. It's a pretty new thing, isn't it, relatively speaking?

DL: Relatively, yes. They were first discovered in Italy, where a laboratory was recording from a cluster of neurons in monkeys' brains. There was a monkey who was just sitting aside waiting his turn, watching another monkey reach for a banana and then peel it and eat it. And a clever technician noticed the cell recordings from this monkey and that his motor cortex was going crazy—the part of his brain that would be active if he were actually reaching for something and peeling it back. They thought this was strange. Do we have our wires crossed? You know, we're measuring this monkey's brain and not the other. They looked into all possible explanations.

They eventually replicated it with a number of different things, and it turned out that they had discovered what are now called, loosely, mirror neurons: neurons that mirror the activity of others. It's sort of the old monkey see, monkey do. So then the question is, How does that happen? How is it that monkeys learn to imitate behavior?

One of the great mysteries in human behavior was that a newborn child can look up at its parent, and the parent smiles, and the newborn will smile. Well, how does it know how to do that? How does it know by looking at an upturned mouth what muscles it needs to move to make its own mouth turn up? How does it know that it's going to produce the same effect? There's a whole complicated chain of neuroscientific puzzles attached to this question.

DB: So when you watch a performance, sports for example, you're not only watching somebody else do it. In a neurological kind of way, you're experiencing it.

DL: Yeah, exactly. And when you see a musician, especially if you're a musician yourself—

DB: Air guitar.

DL: Air guitar, right! And you can't turn it off—it's without your conscious awareness. So mirror neurons seem to have played a very important role in the evolution of the species because we can learn by watching, rather than having to actually figure it out step-by-step.

DB: Yeah, and not only that. You also empathize, you feel what they're feeling. You perceive this person doing this thing, whether it's singing or making music or performing, and the emotions that they appear to be going through and expressing nonverbally. And you immediately empathize with them, and you start feeling the same thing that they appear to be feeling.

DL: Yeah, and I think, ultimately, some aestheticians and philosophers would say that the goal of art is to get you in the same mindset or heart-set as the artist was in when they created the work. They're trying to create a mirror emotion experience.

Stevie Wonder told me that he wrote songs by putting himself in a particular emotional state, recalling a specific event or feeling. And then when he recorded them, he tried to get back into the same emotional state.

DB: I would argue that if the song is written well, you don't have to begin the performance of the song in that emotional state. But by the time you get to the end, the song will have regenerated the emotions that you want to express. So you end up with the feeling that you want to express, but you don't have to have it going in.

DL: Right. And there is a neurological basis for this, actually. It starts with the finding that when we're imagining music, it uses the same neurons and circuits as when we're actually hearing it. They're almost indistinguishable.

So when you're imagining or remembering something, it could be music or a painting or a kiss, disparate neurons from different parts of your brain get together in the same configuration they were in when you experienced it the first time. They're members of a unique set of neurons that experienced that first kiss or that first bungee jump or whatever it is that you're recalling.

Actually, it's in the word "remember"—you're re-membering them. You're making them members of this original set again. I think that's what memory is.

DB: I agree that it's something like a network or a cloud.

DL: Yeah, it doesn't exist in one place. It exists everywhere.

DB: Okay, so someone like Steven Pinker might say that music, while it might be pleasurable, is basically an evolutionary byproduct, a side effect of something else.

DL: Yeah, that's his story and he's sticking to it.

DB: He's still sticking to it.

DL: I saw him a couple of weeks ago, and he's still sticking to it.

DB: Right now I'm pretty much won over by the other point of view, that music and art and other kinds of "nonuseful" skills that we have, that don't seem practical, at least on the surface, do have a use. Music, for example, is good for you and it's healthy. You'll live longer if you listen to it and you enjoy it.

DL: Yeah. Well, the Pinker argument is that in spite of the fact that we find music pleasurable, and it can prolong life and we devote a lot of our energy to it, it wasn't an evolutionary adaptation. Language was the adaptation, and music is sort of piggybacking on it.

He draws some analogies. For example, from what we know, birds didn't actually evolve feathers in order to fly; they evolved them to keep warm in the climates and environments they were in. Once they had this feathery stuff, they then later co-opted it for flight.

DB: Yeah.

Daniel Levitin and David Byrne

DL: There's no evidence that the purpose of feathers was for flight. I mean purpose metaphorically, because, of course, evolution doesn't have a purpose.

Pinker's argument is that humans didn't evolve music for a purpose. Once we had language, we exploited the language that was there, as birds exploited feathers. I don't know if, at the end of the day, the argument really matters, but I think that Pinker is wrong because, for example, in very, very primitive structures that all reptiles have, that all vertebrates have, including humans, there are projections from the ear to the cerebellum and to the limbic system. And these projections convey music almost selectively as opposed to language. Which suggests that music might be evolutionarily older than language.

Also, I mean, we're talking about how music and art seem to be able to convey things that language can't. Well, why would that be?

DB: Mm-hmm.

DL: So when you ask a question like that, you're coming up against the evolutionary-origin question. I've been thinking about this, actually, and it doesn't surprise me that we use art and music to communicate so many things that language won't. What surprises me is that we're able to get as far as we can with language.

Describing something using language, I think, moves it further away from the actual experience. We resort to dance, visual art, music, and lovemaking as a way to express things that are not expressible in language. I think that those are the primary forms of communication, and language is secondary.

DB: I'd agree. Although I'd say that there's an aspect of language that's musical, that has those qualities that lyrics of a song or a poem has, the melodic ups and downs of where the vowels and the consonants hit. There's an emotional content to the pure sound.

DL: Yeah, and to the quality of the person's voice.

DB: Yes.

DB: I once toured with a musician who has perfect pitch. We were traveling on the bus, he had a little keyboard and he would hear car horns and he knew exactly what note it was. He knew it was C sharp, and he could then start improvising something in the key of the traffic. It was beautiful and it was funny and all that. But it made you start to wonder, okay, when is it noise, when is it traffic, and when is it music? Because for somebody like him, the line gets crossed. He finds it annoying sometimes because the car horn will be a C sharp, but an out-of-tune C sharp.

DL: Firing off an angry letter to Ford.

DB: And then there are those who have synesthesia, where sounds generate colors, or worse, tastes. So a sound will actually physically taste sour to them.

DL: Yeah, but that's interesting because when it's an extreme form, they can't tell whether they're tasting or hearing something.

DB: Really?

DL: Yeah.

DB: They don't know which sense is the trigger?

DL: Right. There's actually a theory that all infants are synesthetes, and that sensory differentiation takes a few months after birth to occur. And that infants live in this sort of psychedelic world of everything being jumbled together.

DB: Wow.

DL: But for most of us, the studies from my laboratories and others have shown that language, environment noise, and music all have separate cortical representations—they register and show up in distinct parts of the brain.

And then there are things like what I would call paralinguistic noises. Things that aren't language but that are associated with the expression of the human voice: laughter, crying, sneezing, cough-

Daniel Levitin and David Byrne

ing, hiccups, groaning. These things seem to have their own representation but are closer to the language sections. But we do make a distinction neurologically between speech, music, and environmental sounds.

DB: Right.

DL: When we sing lyrics, both mechanisms, both sets of structures, are being activated. And if you look at, for example, the acousmatic composers from Belgium, France, and Quebec who create entire musical pieces out of jackhammers and waterfalls, there is this sense of ambiguity where your brain is recognizing the sounds as environmental sounds, but the music part of the brain is getting activated, too.

In Pink Floyd's "Money"—maybe the first popular recording that did this—cash-register noises make music. And the brain, I think, responds to that with both mechanisms, which means more of your brain is actually reacting.

DB: Right. Generally, we don't see them as music, but if they fall into a pattern—

DL: Or once a composer places them in a pattern, then you get it.

DB: Somebody orchestrates a bunch of car horns, and it's music.

DL: I've got a recording from a guy named Woody Phillips, who is a carpenter. He noticed that if he put particularly dense wood through his power saw, it would slow it down a predictable amount. So, normally, the power saw would be like *zzzz*, but if he pushed through a two-by-four of pine, it goes *bzzz* and a two-by-four of maple goes *bzut*. So he lined up pieces of wood, and he performed a piece of music—Beethoven's Fifth.

DB: Oh—ha!

DL: I'll send it to you. Any one of the sounds in isolation you wouldn't associate with pitch, but you hear the pitch difference as the saw runs through them. Think about how extraordinary that is. The brain is able to map something it's never heard before. It's never heard Beethoven's Fifth played on power tools, but it knows. We don't have a computer that can do that. The most powerful computer at NORAD, or wherever, can't take a power-tool version of Beethoven's Fifth and tell you what song it is.

DB: Yeah.

DL: It can't compare the live version of "And She Was" with the studio version and tell you it's the same song, but the brain does that—

DB: Immediately.

DL: In seconds. It recognizes the pattern. The brain's looking for order and form. It's a fabulous pattern detector.

DB: Yeah.

DL: It's why dots in a Monet painting become trees, when, really, when you look close they're just dots. Or fusilli becomes a chair.

DB: Yes.

DL: The chair in your art exhibit, that's fusilli, isn't it?

DB: Actually, it's macaroni.

DL: So you're interested in trance states.

DB: Yeah, I am. I'm interested for a few reasons. One, because there's a lot of popular secular music that I think borrows from sacred music. And because of the way it generates a kind of trance state or a transcendent state in the listener.

DL: Yeah.

DB: So you see through the crack in the door or whatever—you can see that wow, this music is taking me to a place that generates all those kinds of vaguely spiritual feelings—like I've gone outside myself, or I had an out-of-body experience, all these kinds of things. And music is often talked about in these spiritual terms. So I feel that there's something going on here. Obviously, these musical experiences are touching another part of the brain that's linked to a kind of spiritual or religious experience. Probably because it takes us out of ourselves in that kind of sense, for want of some better term.

DL: I think that's a perfect description of it: out of ourselves.

DB: And it's a little bit of ego loss, which, like in Eastern philosophy, is a kind of pleasurable, transcendental experience. You become one with all these other people around you.

DL: Yeah.

DB: So it's like all of a sudden you're part of the hive; it's this wonderful feeling.

DL: Right, and this isn't unique to music. There are other things that will get you there, chanting, breathing, et cetera.

DB: Yes.

DL: And I think what they all have in common—you nailed it when you talked about being out of yourself—is the sense that your thoughts are not under your conscious control, that something else is guiding them, though you're still aware of them and can bear witness to them.

That's what dreams are, in a sense. And why music or rhythm is able to induce this state is a mystery. Nobody knows. We do know a little bit, neurologically, about what's happening. We know there's a suppression of frontal-lobe activity. We can measure changes in alpha waves and gamma waves and things like this. But those are really descriptions, not explanations.

DB: Uh-huh.

DL: We don't really have the ability to explain how it happens or why. But it does seem to have something to do, if you'll let me speculate, with this balance between seeking order and predictability and violating that order and predictability. And when you have a complex pattern of rhythm or pitch, which is what music is, you relinquish some of your control. You're in a state of relaxation, you're following along this stream of sounds. You're making yourself vulnerable, giving in to the music. And you're lulled into this state of half sleep, half wakefulness, is the best way I would describe it.

It's a powerful experience to have with other people. Which is why in the sixties, when people took drugs, the classic thing was sitting around on pillows with a group of people. Or when you go to a concert with twenty thousand people, to some extent you're out of yourself, but you're with everybody else and it creates a bond. I think the Grateful Dead had this.

DB: Yeah, it's not strictly a musical experience.

DL: Right. And I really think that one of the ancient functions of music was for social bonding. I mean, there's a lot of evidence. If you go to current tribal communities that have been cut off from Western civilization, they use music to form community.

DB: Mm-hmm.

DL: Each tribe has its own music. The men stay up to ward off predators by singing around the campfire. Music is communal. It's almost ironic that today technology and culture have taken us to where we all have our little ear buds and we listen to music in private, given that for tens of thousands of years, the only way music was experienced by humanity was communally. Everyone played music with each other. There wasn't a separate audience and performer. And dancing was always a part of music making. It was a big communal activity.

I think our nature, as you were pointing out with the mirror neurons, is to move when we hear music. To move with people and to have it be part of a group experience.

DB: Right.

DL: I'm working with a group of people with Williams syndrome. They have a genetic abnormality and stand in contrast to people with autism who are very shy and antisocial, and they don't seem to really understand music. People with Williams syndrome are hypersocial and hypermusical. So you've got these cases where music and sociability go together in Williams syndrome, and they fall apart in autism. It really suggests that music and sociability are part and parcel of the same thing, communication.

DB: Yeah. When I began making music professionally, I was an extremely shy person. Very kind of socially inhibited. I feel now, I'm just guessing, but I would say I had—

DL: Asperger's?

DB: Asperger's, yes absolutely. I think probably that's what it was.

DL: I feel that about myself sometimes.

DB: But, you know, I think I used music—I mean, maybe I just outgrew it, because I think that happens to some extent—but I also think I used music to find a way into engaging socially. I thought of it really like a hammer or a pair of pliers. It was a tool.

DL: I think what many listeners get from you is that you were a shy guy, a somewhat socially awkward guy, trying to break through. And they identify with that because everybody feels socially awkward at one time or another even if they're homecoming queen or class president. I think the strength of your musicianship and artistry was that you were able to lay that out in the open in such a way that people saw your struggle to be social. It came out in your music and your stage presence and the persona that is David Byrne. I think that's a powerful thing.

DB: And then I wondered, so if I heal myself what do I have to say? What's left to do? Never mind that's another whole thing.

DL: You probably don't feel 100 percent.

DB: No. I don't feel that. There's always stuff to do and there's always stuff to say and there are always things that engage me, that get me excited.

DL: I'm so far from healing myself that I don't even entertain that thought! ∞

Eleven

On Shape

Lisa Randall and Chuck Hoberman

The physicist and the inventor meet up to talk about geometry, creativity, and the shape of the universe.

Lisa Randall first saw Chuck Hoberman's work while on a ski trip in Utah in 2006. His famous Hoberman Arch, which opens in the same way an eye dilates, is the centerpiece of Salt Lake City's Olympic Medals Plaza. Hoberman's unique use of shape, scale, and dimension in his transformable designs seemed richly analogous with Randall's use of extra dimensions and warped geometries in her research on the nature of the universe. *Seed* invited them to explore this conceptually common ground.

Harvard theoretical physicist Lisa Randall is a leader in the fields of particle physics and cosmology, and is especially renowned for her work on extra dimensions. In 2005, the *New York Times* included her book, *Warped Passages: Unraveling the Mysteries of the Universe's Hidden Dimensions*, in its 100 Notable Books of the Year. Randall was also included in the 2007 Time 100, *Time*'s list of the most influential people in the world, and *Esquire*'s 2008 list of "The 75 Most Influential People of the 21st Century."

Chuck Hoberman is perhaps best known for inventing the Hoberman sphere, a geodesic globe that can expand up to five times its diameter. An accomplished designer, architect, artist, and engineer, Hoberman has pioneered the field of transformation technology in service of such diverse areas as medicine, architecture, and toymaking. He won the Chrysler Design Award in 1997 and his creations have been displayed around the world, including at the Museum of Modern Art in New York and the Centre Pompidou in Paris.

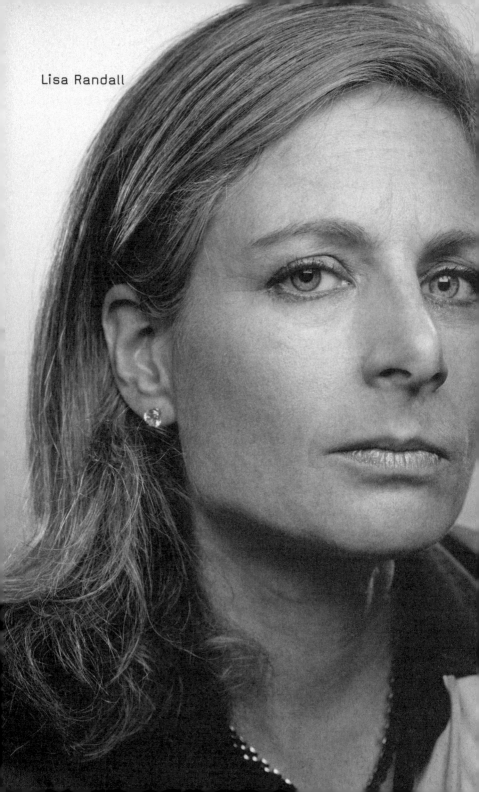

Lisa Randall

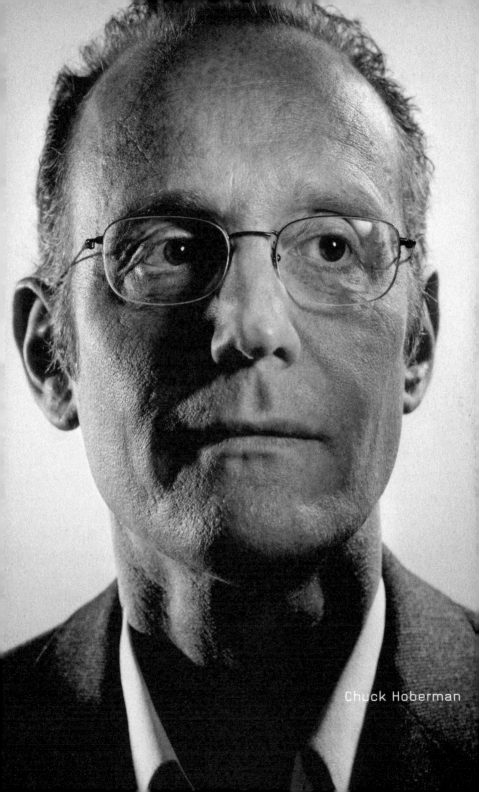

Chuck Hoberman

Chuck Hoberman: You present yourself as a "model builder" in your book. That was something that stuck with me. You're a scientist doing advanced physics, but is there some aspect of your work where model building is really a kind of design?

Lisa Randall: That's an interesting way of thinking about it. It's funny, because when we design physics models we'll often talk about it in those words—designing models or creating models. But what we're really doing is trying to reproduce reality by guessing what's out there and searching for ways to test those hypotheses. We're asking, "What are the underlying design principles that actually exist in nature?"

CH: Right. What does "reproduce" mean in that sense? Because you're not reproducing it, you're making a model.

LR: Well, in science, we make certain assumptions about what the relevant elements are in order to make predictions that match what we're able to observe. So, if we identify the correct starting point—the correct ingredients, the correct laws, or forces—then we should be able to reproduce relationships that would perhaps otherwise be mysterious. We take theoretical ideas, work out their consequences, and then see if the consequences are observable.

CH: So that would be a key criterion for model building—you want a model that has close-term predictive consequences?

LR: That's right. One of the exciting prospects for the future of physics is the Large Hadron Collider. It's going to collide together protons at higher energies than we've achieved before. In the process, it's going to test ideas about how particles get mass, and why gravity is as weak as it is. Right now all we have are models to answer those questions. We know the mechanism, something called the Higgs mechanism, that explains how particles get mass. But we haven't observed the underlying reality, so we don't know which, if any, of these models is correct.

But let me ask you—when you build things, how abstract are your ideas in the beginning when you first start your designs? For example, Einstein is well known for thought experiments. I'd say that those are abstract, but they are still very much tied to physical things that you can picture, like riding a beam of light.

CH: Right.

LR: So, when you conceive of your designs, do you have something physical in mind? Are you thinking in concrete physical terms or are you going beyond that?

CH: My engineering training is in kinematics and mechanisms. And a mechanism isn't defined by the physical pieces that make it up; it's defined by the relationships and connections between the pieces and the trajectories. So, in that sense, the thinking is quite abstract. But there is a very strong visual aspect—not pictures of a thing, but images of processes and relationships.

Can you picture a brane? Does it have a shape or is it simply that you posit a brane as something purely living in an equation world?

LR: It absolutely has a shape. It has geometry. It has a geometry that you can determine through working out Einstein's equations and figuring out what the energy distribution is and figuring out what the particular boundary conditions are. If we live on a brane, it would have to look like the observed universe, which is big and flat.

CH: This is something I think about a lot—the relation between visible shapes and their underlying mathematics. For example, in designing mechanisms, a very important concept is degrees of freedom. You can do a simple equation to determine if a mechanism is a fixed object, a smoothly movable object, or a "floppy" object—one that has multiple degrees of freedom. This equation gives you a single number that is associated with a particular object and that is a predictor of that object's behavior. But the equations will tell you only so much.

LR: Can you give me an example?

CH: Sure. A four bar linkage—four members that are connected by four pivots—has one degree of freedom. In other words, it has one variable that describes its state, for example, an angle between two links. Then, when you make something that has, say, a thousand pieces, you can calculate its degree of freedom, but the equation doesn't tell you quite the same thing. In that case, the single degree of freedom predicted by the equation—a thousand pieces moving synchronously—would, in fact, be a floppy mess. Why? Because you haven't accounted for the flexibility in the pieces themselves.

LR: So there are degrees of freedom that haven't gone into it yet.

CH: The equations aren't quite describing it, but it sparks a thought process of how to come up with criteria by which you can pursue these more complex mechanisms and still predict their behavior.

LR: Is it really criteria or is it just that you haven't put in all the relevant degrees of freedom?

CH: It's probably the latter. For someone like me there's no point in analyzing or modeling past a certain point. But that's the difference between engineering and science.

LR: Right. So, is that how things work out for you that you accidentally find that there are degrees of freedom that you don't anticipate?

CH: No, it's not accidental. I may design a mechanism that the equations will say is overconstrained and should be a fixed structure, because there are too many connections relative to the number of pieces. But actually it moves smoothly because of symmetries within the underlying geometry of the structure. These symmetries reveal themselves as so-called invariants in the system, say as angular relationships that are unchanging even as the mechanism moves. It is precisely these extra connections that give the mechanism its structural integrity, so that it can perform two functions—it can hold its shape and form, and yet it can move in a precisely synchronous matter. Normally I don't actually describe my work that way but that is in fact the way I think about it.

LR: That's interesting and in some respects relates to one of the key features distinguishing creativity in science from other forms of creativity, which is the constraint that, ultimately, your models have to match reality.

CH: Which is the ultimate constraint, in a certain way.

LR: Yes. In terms of the actual doing of science, it's a creative challenge to figure out which problems you might ultimately be able to solve. In part it has to do with what you were talking about, with respect to the number of degrees of freedom: how many measurements you can make versus how many inputs you have to give. A lot of the time it's a question of building on what you had before, so that you have all the degrees of freedom you had before, and then you postulate, maybe, there's one or two more needed to more fully describe your system. Or maybe there's some underlying symmetry, or force, that you hadn't anticipated. So you try to make models in a way that you're not introducing more additional structure than you need to so that you can make it as predictive as possible. You try to build a more encompassing theory that agrees with all known measurements.

CH: Right.

LR: Model building in particle physics has other constraints built in as well. Because of quantum mechanics, any interaction that could possibly occur, will occur. So you might say, "I'm introducing only a few interactions," but you find out you've introduced a mess of them.

CH: Is that the "anarchic principle"?

LR: Yes it is, but that's a term I introduced in my book motivated by suggestions of the physicists Murray Gell-Mann and Jonathan Flynn.

CH: I love it. All those little particles jumping around.

LR: Yeah. But these "virtual" particles allowed by quantum mechanics tell us that you can't necessarily say, "I'm going to introduce only three new variables." You often find more that tag along whether you wanted them or not. That's one of the things that came up in a theory called supersymmetry. People thought it was a simple thing—we're introducing a few new parameters—and then realized there really are actually many more parameters than were desired. Which then introduces questions like "Which interactions are really permissible?" and "Can I constrain those?"

CH: And which, actually, has virtually a one-to-one analogy with the design process, in the sense that just when you think you've hit upon exactly the right solution, the law of unintended consequences jumps in, and you find that you've opened up a whole new can of worms.

LR: Right. So there are elements or motions that you hadn't anticipated.

CH: Exactly—from a functional standpoint I've solved one thing, but I re-jiggered the design and now something else may not work.

LR: What's driving you? How do you decide what you want to make?

CH: Well, because I have a business, if a client asks me, I respond. Actually, it's an interesting way to pursue what ultimately is an idea-based practice.

LR: What do you think people find appealing about your designs?

CH: A mathematician once told me that "the appeal is what mathematicians enjoy about math that they can't convey to people who don't do math." So, when people say "that's kind of mesmerizing," they're seeing something that is quite visceral, but they're perceiving relationships as opposed to things.

LR: One of the challenges in communicating physics—and obviously it's not something big and sculptural that people can look at—is actually being able to express what questions are driving us.

CH: Right.

LR: One of the things that drives us is what you were talking about. We like fitting things together, we like puzzles, and we like seeing consistency. But the reason we're physicists and not just puzzle-solvers is that

Lisa Randall and Chuck Hoberman

we do ultimately have some underlying questions and we would like to see them latch on to reality.

It's kind of interesting to think about motivation or appeal in terms of design and sculpture. Because it sounds like, for you, one of the appeals is being able to make things work and fit together, but the appeal to the onlooker can be quite different.

CH: Unfortunately, one difference is that while everyone else is going "ooh!" and "ah!" at the sculpture, I'm thinking, "Gosh, I hope the cable isn't wearing out." Because I know too much and I can't really see it the way other people see it.

LR: That's always a fascinating feature of any creative work. The person who created it often sees it so differently from the outside observer.

CH: In science, the concept of elegance or aesthetics often comes up as something desirable. Perhaps you wouldn't want reality to be inelegant—it would be a disappointment. I'm wondering if that's still an important idea?

LR: You know, it's really interesting because I thought about this quite a bit when I was writing my book. For example, string theory is often described in terms of elegance. But for the kind of modeling work I do, one of the things to understand is how symmetries can be broken in a compelling way that still retains elegance. You don't want something that looks like you really had to just jimmy everything through some sort of Rube Goldberg machine. You want something that is elegant and explains what we actually see. I mean if everything were symmetric, it wouldn't be as challenging a problem, right? But the fact is, most of the symmetries are broken when we look at the universe around us. The question is, How can the underlying thing be symmetric and still yield what we observe today?

CH: How do you describe broken symmetries?

LR: There are different types of symmetry-breaking. One known as spontaneous symmetry breaking can probably be best explained through examples. One I give in my book is that you're seated around a dinner table on which are placed wineglasses to everyone's right and left. The table is completely left-right symmetric. But, as soon as someone chooses one of the glasses, say the right-hand glass, everyone's going to choose their right-hand glass, and then the symmetry is broken. In this example, the symmetry is broken by the actual state of the system, but not by the underlying laws.

CH: Right.

LR: Another thing is that there's a scale associated with the symmetry breaking. In other words, at high temperatures, for example, the universe manifests symmetries for which the symmetry breaking is noticeable only at low temperatures.

CH: It's interesting, symmetry has a different connotation in terms of design. I give a lot of talks to architects and architecture students. They'll often look at my work, which can have a kind of mathematical regularity to it, and they'll say, "Oh, it looks so symmetric." And they're quite disappointed.

LR: The fact is most things in the physical world aren't the most beautiful when they're completely symmetric. It's sort of the small breaking of symmetry that is always intriguing, I think. So I'm curious what people say about your things being symmetric.

CH: Well, the way that an architecture student thinks about it is pretty much in terms of the look of a thing. And they may feel that it should look different from what came before. So I think there's a sense that symmetry is old-fashioned in terms of a style.

The point which I try to make to them is that symmetry is never an absolute. There are degrees of symmetry. A glass, for instance, is rotationally symmetric, but in other ways, it's not symmetric.

LR: Right—a glass respects some symmetries and breaks other symmetries.

CH: Precisely. It's a more complex situation than to look at a particular artifact and say it is symmetric versus asymmetric.

LR: When you build something that's completely symmetric, do you ever have an instinct to make it asymmetric, to just wonder . . . what if that piece was out of place?

CH: Well, you know, I'm not really primarily motivated in my designs by the way things look. Their look emerges out of other concerns and constraints. I try to design things not to look a particular way, but to perform in a particular way.

LR: So, when you look at it, is that what you're seeing? You're seeing the performance?

CH: Yeah, that's much more my focus.

CH: One area that we work a lot in, especially for our product designs, is foldability— making furniture or tents or even toys that fold down small.

LR: People love that, too. In addition to the actual practical element, for some reason, seeing things that can fold up or become compact is something people respond to.

CH: My sense is that people's perception of an object that expands is closely tied into physiological sensations of organic growth. After all, we started small and then we got big, so the scaling of an object becomes associated with a living quality. In that sense, my approach is a form of bio-mimicry, the close connection between the human-made and the natural, which I think has a very basic appeal.

LR: Yeah. Also, the idea that solid material is mostly empty space is rather mysterious to most people. Yet the idea that you can have something big fold up into something small—which means implicitly that the big thing has to be mostly empty—is so readily understood and accepted.

CH: Oh, one hundred percent, yeah.

LR: Why is that less confusing?

CH: I think because you hold it in your hand. The ability of my sphere to expand and contract may be slightly mysterious to people, but its tactility gives a sense of familiarity. I think one of the reasons people enjoy our toys is that it's like a magic trick, but there's nothing hidden. So one might say, "Oh, I see everything! But wait a second. Why does it work?" And then you scratch your head . . .

LR: Right.

CH: I could explain to you simply and precisely why it works but it's a bit like you explaining your physics. If you're not into trigonometry, even after the explanation, the question just gets repearted: "Yes, but why does it work?"

LR: And then, can they go off and design one of their own?

CH: Well, that's interesting. As a toy designer, I always hoped to inspire another generation of folding mavens. And pretty regularly I'll get some eleven-year-old geometry whiz sending me fantastic origami bits and all of that.

LR: Really? That was you when you were young, I bet. How did you get into making transformable objects?

CH: I originally studied art, and I made kinetic sculptures, which sometimes didn't work very well—this was my motivation to go to engineering school. Around the time I graduated I wanted to find an art problem to focus on. This was the mid-eighties, and personal computers were still pretty new, so I was turning over the idea of pixelization and the thought came: You know, maybe I should build a sculpture that will have three-dimensional pixels. The next thought was "Okay, I need each pixel to be able to turn on and off. When it's on, it's visible, and when it's off, it's invisible. So I have to make something that can disappear." And then, "Well, I can't really make it disappear but what I can do is make something big get small." That was my framing problem, if you will. I got quite involved with this notion of a three-dimensional pixel that could appear and disappear, and eventually forgot about the sculpture itself.

LR: That sounds like quite an interesting idea.

CH: It was a good starting point. That magic of appearance and disappearance is still very much a pursuit in my designs.

LR: Yeah, it's funny. One of the things that underlies so much of physics is the notion that literally seeing is only one way of detecting things. I'm always trying to find ways to, in some sense, find the invisible.

CH: So what does that mean? That you see it in your mind's eye, or that you have a familiarity on some other level?

LR: From a scientific point of view, seeing means detecting. So you may observe it at a nonvisible wavelength of light but it's still detected because a detecting device is recording it.

CH: Okay.

LR: So in a sense some forms of "seeing" scientifically are less intuitive. I often get asked about this because for some people that would be less real, less authentic.

CH: Right.

LR: For example, I will say we see quarks, and people ask, "You don't really see quarks, do you?" And the fact is, we observe the experimental evidence of the existence of quarks. If it's a heavier quark you will detect what it decays into; if it's a lighter quark you will see that it's a strongly interacting object, you will see that it has energy, you will observe that it has interaction cross-sections, you will observe that it has all the properties you say a quark should have.

CH: So it is simply mediated in a way that directs your eye to something.

LR: But there is a little bit of theory. You have to understand the theory to understand what the consequence will be. For some people that leap is so indirect that they will say you're not really seeing it. For us, as scientists, we are seeing it.

CH: But of course in a certain way science is just making explicit what we all do. . . . The way we make sense of the world is because we have models of the world. I mean, right? That is what brain science tells us.

LR: In fact, we have many reasons, as you know, to not always directly trust our senses. I mean there are all sorts of optical illusions and all sorts of ways people have shown that we don't actually always see exactly what is there. It depends on how we are looking. So it's nice that there are more precise ways of recording it and actually being able to ascertain what is really there. ∞

Twelve

On Artifacts
Michael Shanks and Lynn Hershman Leeson
The archaeologist and the artist meet up to talk about presence.

In 2005, Michael Shanks and three colleagues started The Presence Project to explore issues of presence and documentation across the arts and sciences. Lynn Hershman Leeson joined soon after and, together with Shanks and others in the Stanford Humanities Lab, created *Life to the Second Power*, an online encounter with her archive. As they see the project through to its completion in 2010, Shanks and Hershman Leeson plan to further explore memory, identity, and place. *Seed* invited them to advance the conversation.

Michael Shanks is the Omar and Althea Hoskins Professor of Classical Archaeology at Stanford University, where he co-directs the Stanford Humanities Lab. He also directs the Archaeology Center's Metamedia Lab, which explores the materiality of media and the role of history and memory in art and artifact. Shanks is the author or co-author of several books, including *Re-Constructing Archaeology*, *Experiencing the Past*, and *Theatre/Archaeology*.

Lynn Hershman Leeson is an artist whose works span photography, film, video, performance, and installation. Her artwork has been shown at more than two hundred major institutions and is part of the permanent collection at New York's Museum of Modern Art. Hershman Leeson is professor emeritus at the University of California, Davis; A. D. White Professor at Large at Cornell; and chair of the Film Department at the San Francisco Art Institute. In 2009, she was awarded the John Simon Guggenheim Memorial Foundation Fellowship for her forthcoming documentary, *!Women Art Revolution A (Formerly) Secret History*.

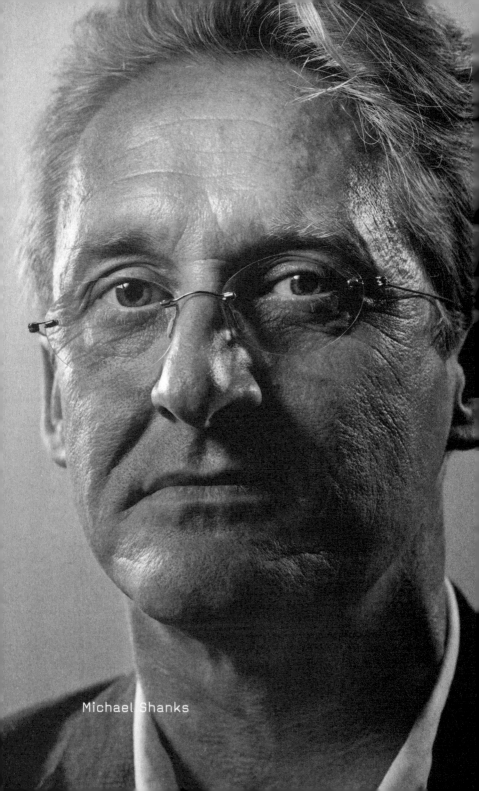

Michael Shanks

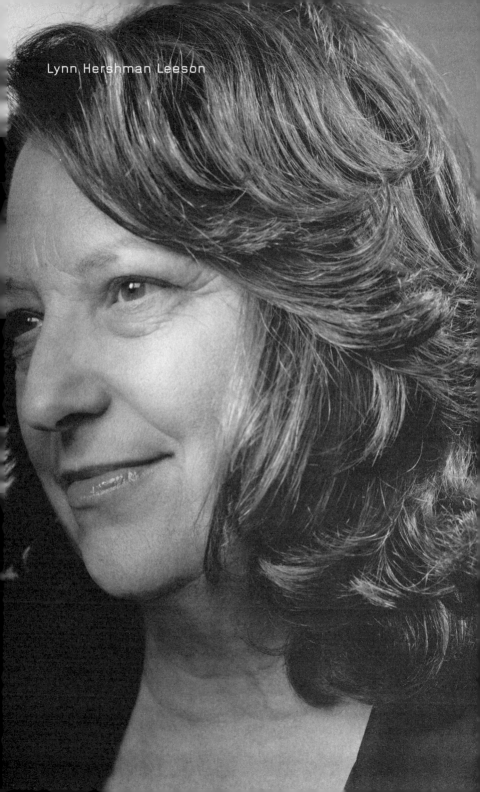
Lynn Hershman Leeson

Michael Shanks: 1972: You were working in San Francisco, and you did a piece at the Dante Hotel.

Lynn Hershman Leeson: Yes. I'd done a piece with sound at the museum, but they said media wasn't art and didn't belong in an art museum. So I thought, "Well, why not just use an environment, wherever it exists?"

So Eleanor Coppola and I created rooms in the Dante Hotel, which was a run-down place in North Beach. It was very simple. We rented the rooms—mine was rented indefinitely; hers was rented for two weeks. And I created a situation where people could look at presumed identities constructed from artifacts placed in the room.

MS: So you put stuff in there?

LHL: Yeah. I put goldfish in there. There was a soliloquy of Molly Bloom. There were books that the presumed people might have read, clothing that they might have worn. People were invited to trespass. It was open twenty-four hours a day; people could check in at the front desk, get the key, stay as long as they wanted, and displace it.

MS: Did anybody leave anything behind?

LHL: Nobody left anything. They graffitied the mirror that was there, but nobody took anything. They really respected that space.

MS: Were you monitoring people coming and going?

LHL: No, not really. It was left gathering dust and the flux of time as people traveled through. I was just starting to think of time and space as elements of sculpture at that point.

MS: And then the police came at some point, didn't they?

LHL: Yeah—ha! Somebody reported a body in the bed, because there were these wax cast figures—

MS: Which had been there from the beginning?

LHL: Yes. And the police confiscated everything in the room and took all the artifacts down to central headquarters, which, I thought, was really the apt ending to that particular narrative.

MS: And then thirty-two years later, Stanford acquires your archive of ninety-something boxes. The remains of your body of work—whatever hadn't been taken away by the police, I guess!

LHL: Yes.

MS: As an archaeologist, I'm interested in what comes after the event, as it were. What you do with the remains of the past, to somehow try to get back to where they originated.

LHL: I don't know that you can ever get back to that point, but you can go forward, using them as context for the future. The trail and the remains may be dormant, but they exist, waiting to be revived or resurrected into something else.

MS: Yeah, regenerated. This is one of our major points of contact. A lot of people think that archaeology—archaeologists—discover the past. And that's only a tiny bit true. I think it's more accurate to say that they work on what remains. That may sometimes involve, absolutely, coming across stuff from the past—maybe a trilobite fossil, or a piece of Roman pottery, or, as my colleague Henry Lowood and I did, your boxes in the Stanford collection—but the key thing about archaeology is that it works on what's left. And that makes of all of us, really, a kind of archaeologist. We're all archaeologists now, working on what's left of the past.

And you're right, as we explore this stuff, we figure out how to bring it forward, first into the present, through our interpretation of it.

LHL: Exactly. I didn't want the work to remain in boxes. Much as I love the Stanford Library and Special Collections, I wanted this to be more universally accessible. I suggested to Henry that, possibly, we could make a game, a mystery, or a film noir about the remains of this evidence of a life, which portrayed itself in various episodes. Henry suggested a possible adaptation into Second Life, which then became the *Life to the Second Power* project.

MS: And it connects with the interest that we share in the nature of the archive. Boxes, in a collection, vitrines in a museum, they're often—and appropriately—seen as quite static.

LHL: That's right. Static but charged.

MS: Unless there's a reason to reuse stuff, it'll fall out of use or be stored away; and, eventually, it'll end up in a landfill site, if you're lucky, or destroyed. So the question we share is how to reanimate the archive.

LHL: Exactly. Revitalize the past, inserting it into the present, which gives direction to its future.

MS: Yeah. Displacement is another key feature of this archaeological sensibility. What happens when old stuff—remains—are shifted into new associations.

LHL: And it's particularly interesting because Second Life and some of these social-network programs involve notions of trespass that have no geographic boundaries. So it's taking the exact same premise of this project, the Dante Hotel, from thirty years earlier and transplanting it into something that allows a completely different, but yet related, experience.

MS: Yeah. There are parts of our contemporary attitude toward spaces and places that are very archaeological. It's about how we almost automatically and subconsciously look at spaces in terms of evidence. It's a forensic sensibility.

Archaeologists survey and excavate places. They document, map, collect, and categorize, seeking to identify what generated the remains—for example, past events, social or environmental changes. Archaeological evidence is thus treated as symptomatic traces of deep structures or events, archaeology is a hybrid science of material traces.

The detective, another nineteenth-century invention, also connects evidence with event and place. But how do you know what might be the key evidence at the scene of a crime or an archaeological site? Anything might be relevant. Anywhere could be the scene of a crime. This is what I mean by forensic sensibility. Anything could be the trace of something that once happened there.

LHL: But now, with the forensic sensibility, there's also a digital demeanor that didn't exist before.

MS: Oh, right. "Digital demeanor," I like that.

LHL: A digital demeanor of trespass using interaction to reveal the evidence.

MS: Yeah, which brings up implications for storage, for retrieval, and, of course, surveillance, looking, watching, and how these have become incorporated in all sorts of digital technologies.

For two centuries and more, archaeologists have been developing a tool kit for working upon the traces of the past. They're concerned with a kind of genealogy—*how* the past, in its traces, has come down to the present, rather than the traditional sense of history as what happened *in* the past.

LHL: This can involve the trauma of memory.

MS: Oh, yes. This is absolutely archaeological. We often feel separate from the past, and then, in that separation, we visit a room, such as the Dante, and we instinctively look to piece together what we see in front of us.

Again, working on what remains.

LHL: But, in the particular case of our project, Life Squared, you're able to see the evidence being looked at and to lurk inside and watch somebody else discovering the evidence and re-create endless narratives, as they repattern the same information and create yet another trail of how it's being seen, reseen, recomposed, remixed, so that there are an infinite number of ways you can perceive it.

MS: And, I think, our digital demeanor, as you put it, precisely foregrounds us again. I mean I could argue that it's always been a component of what we do: taking up bits of the past, reusing them, reworking them, which absolutely implicates issues of memory.

LHL: And erasure of ownership.

MS: Oh, yeah, absolutely. And I'm very keen on countering this notion that, in terms of the past, we need to somehow hang on to it and preserve it.

LHL: How do you preserve it? How do you embalm time?

MS: Well, yeah. In this way, to preserve the past is to kill it off. Transformation, translation is essential if the past is to live.

LHL: Yes.

MS: Just yesterday, I got an e-mail announcing a Web site that essentially comprises virtual reality reconstructions of ancient sites—3-D models of the forum at Rome, or a basilica, or an ancient monument in Greece. These are CAD architectural models visualized in 3-D, so you can walk through them. And they're realistic, in the sense that you can admire the textures, experience the spaces. It's meant to be a very engaging experience of the past—history reconstructed in some kind of photographic verisimilitude—so that it's present to you now. But I find them utterly, utterly empty and dead.

LHL: Why?

MS: Well, to walk through a room in this way, on a computer screen, doesn't necessarily elicit any reaction other than a distracting and superficial one, such as, "Ooh, the texture of the floor is spot on. . . . Ooh, I like the light coming in through that window; it's just right."

And what generates a sense of being there is not this kind of surface authenticity, but the fidelity of narrative. The narrative of these graphics is nothing more than taking a stroll.

These models can be very flashy, highly naturalistic, and look "real" but they don't help us make sense of and understand things—floor plans or the shape of ashlar blocks give little understanding about life in the past. This is the old illusion, that faithfulness to the external appearance of things gives us a hold on reality.

And such models forget about engagement. Not just the experience of visiting old places, but the detective work that turns data into information and then into stories that engage people now.

I sometimes think that these elaborate models of the past are part of a contemporary optimism that a quantitative increase in data will somehow deliver a better understanding of the world. In this kind of digital archaeology I see the dream that eventually, and with so much data at hand, we will be able to relive the past. This is the impossible desire to bring back the dead. I say, look, the past is over and done, decayed, ruined, lost. We only have a few bits to work on. And this is what is fascinating.

Virtual reality archaeology is a project that brings to mind the movie *The Matrix*—the creation of a world that actually doesn't or didn't exist, though it is lived as reality.

LHL: The closer you get to what you think something is, the more evident it becomes that it's also an illusion.

MS: Yes. Absolutely. It's a question of what truly constitutes evidence about who you are, about who I am.

LHL: It's always apparent in the flaws. You know, it's in the crack in the wall, not the replication of it. I mean, that's where the truth is. It hides, waiting to be discovered.

MS: Yeah. It's in the gaps, in the stuff that gets overlooked.

So, anyway, that issue of authenticity, I think, is a big one. Considerable resources, research dollars, and institutional support are being devoted to this kind of VR modeling. And you know, it's . . .

LHL: It's wrong.

MS: Well, it's illusory in *The Matrix* sense. And there's an authenticity there, because all of the stuff that's "left over" is on show, in high-res, so you can zoom in on it and look at it in considerable detail. But by no means is it "the Past."

It's an interesting negotiation between our current means and the ends we have in mind for archaeology. How we document the past connects, obviously, with all sorts of technologies and instruments now. Instrumentalities relating to information, information flow, and organization. The whole field of documenting ourselves is changing as our tools change.

LHL: The information age requires new tools, absolutely. I'm making a piece right now that deals with the five leading blog tags in the world. It's to see what people are thinking about, a global mind-reader. Software reads key words, tags them, and makes "judgments" about the emotional range of information. So it lets us know at a glance the mood of the global mind, as seen in constantly evolving and morphing blogs.

So many things that used to be hidden are now evidenced and present. We're inverting the exoskeleton. For example, there are some wonderful ways to photograph and scan paintings to uncover their histories.

I like to pull forward the things that we've always thought should be invisible and make that a part of the communication structure, in fact, the whole nature of a work. So the invisible becomes the aesthetic itself. Because by revealing process, we reveal meaning.

MS: As you know, I have a deep interest in the history of archaeological approaches to the world, to evidence, to information, to documentation. And it's undoubtedly the case that a lot of this interest in ruination and the interest in decay—the Gothic interest in the dark side of things—is very much an eighteenth-century invention, or preoccupation at least.

It's the idea, the figure, of the undead, of the renegade. It's the perverse count in some ramshackle castle who's coming back to haunt us and thereby, you know, influencing the present.

LHL: Was the creation of the undead simultaneous with the invention of electricity?

MS: Well, certainly it all goes together. There was a barrage in the Age of Reason, the development of experimental methods, of science, of rationality. And this accompanied, of course, a romantic fascination with the other side of reason. The irrational. Whether it's mental or social or cultural. The invention of modern notions of crime comes at this same time. So, deviation, crime, all goes with this hyperrationalized approach to nature and the world. It was about separating the rational from the irrational.

LHL: And deciding which is which.

MS: Yeah and trying to decide between the two, which is connected to another component: the demarcation of what it is to be human. What is human and what *isn't* human.

So, it's the machine and the human, or the inhuman and the human. Or the stuff that is often seen as accoutrement to us, separate from us, whether it's the information that we generate about ourselves, our relationships, or our stuff, our material things.

So, questions of: Is it me? Is it not me? Is this trail I leave in the world around me, this archaeological trace, is it me or is it something secondary? The things I use and own, do they constitute who I am? Or are they just the things that I use? This theme of where do I end and where does the world begin and how am I, as a person, dispersed in the sociocultural world? This is a classic theme that has worried us—in its modern guise—since the eighteenth century. And it goes with the invention of disciplines such as archaeology, anthropology, and human and social sciences.

LHL: But now we're spawning a different kind of mutation, because we're able to reconceive ourselves virally and instantly put that morphed and evolving regeneration into the world specifically so that it can be adapted and changed. So, where does that mutation leave us? Is our sense of presence, and who we are, an appendage to how we are perceived?

MS: I always say that what archaeologists have to make them distinct is the long-term view of things. Absolutely, we're made very conscious of this now. But I see all of this, really, as just coming at the end of a long, long history. I don't think it's new. I think these issues have faced us for as long as we've been human. The phrase that I use is "For as long as we've been human, we've been cyborgs; we've been intimately connected with things, with goods."

In the early days—and I'm going back to 120,000 years ago—I think what made us human was an intimacy with goods, with things, in kind of "machine-ic" assemblages, even though they weren't formal machines.

The temple and imperial administrative bureaucracies of the ancient Near East were what Lewis Mumford called megamachines. They built the pyramids—twenty thousand horsepower running for perhaps six hundred years and capable of positioning a million stone blocks accurate to a fraction of an inch.

LHL: So can autonomous agents even exist, do you think? Or do you think that everything is kind of tempered by these assemblages, this sampling and remixing? How would you determine whether something is independent, isolated? It can't be, in order to function.

MS: Yeah, yeah. I've been fascinated by the way your work explores the limits of what makes someone an authentic self. As an anthropologist I agree that authenticity is not best connected solely to internal properties of an autonomous individual. We find our authentic selves in others and in our relations with goods.

LHL: Everything is defined by its relation to something else.

MS: Right—think of what's happening in this room right now. That is, in the future, looking back, what would be the definitive statement, representation, of the room here and now?

There's a conversation happening between you and me, but even that is influenced by where I've come from, where you've come from, and it will take us in different directions in the future. And, I don't know, maybe in a little bit of time I'll look back on this and say, "Ah, that was when I realized my calling was not archaeology but the arts!" So what happened in this room was—yes, a conversation—but that was coincidental to something else that I realized, only with hindsight, had happened.

But then you might say, what's happening in this room is that the air-conditioning has been switched off and the patterns of heat transfer are now apparent. As a physicist, you have a very different view of things. As an investor, perhaps your perspective is that this is the last use of the building before they redevelop it and turn it into condominiums.

So what's going on here has no bottom line. There's no definitive answer to say, this is what's going on here, and it can therefore be represented in one way.

So the question becomes, How do you take photographs of all of that? How do you make a video out of it all? How would you document it? That is the classic issue, I think: What is the definitive record or representation of something, an event, an occurrence, a person?

There isn't one. Now, this is not disempowering, it's the opposite. It's actually empowering, because it opens the door to actually playing with it; to remixing, reworking the processes of documentation, of engagement, whatever.

LHL: And invisibility, things we can't see now, that are embedded in time, even here in this room, waiting to reveal themselves. And there are many ghosts lurking unseen that it will take generations of inventive science to understand. Our perceptions are limited to the technologies we can access.

So you really can't discard anything. It's only a matter of time before we see what economies determine as being sustainable. It is going to be surprising, not at all what we expect, not at all linear.

MS: I think what you're looking for here, very appropriately, is what I would call the politics of legacy.

LHL: Of presence.

MS: Yes—the politics of presence. What is made present and what is kept absent and invisible.

LHL: But it's never completely invisible, because it can always be traced.

MS: Well, there, again, in my long-term perspective—it's a very melancholic one—I think that most of history is . . . Well, we've just lost it all.

LHL: Ahh.

MS: And I think there is a crucial issue in our current politics, now and for the future, which is what are we able to recall, to document, to trace, and also what should be documented and traced and not kept invisible.

LHL: And who makes those decisions.

MS: Absolutely, it's about power over these processes. It's a crucial issue.

At the same time, as I say, there is a melancholy about our pasts in that so much has been deliberately destroyed or concealed or forgotten. It's the politics of the past. As we all know, it's the winners who write the history books.

But I think, with this digital moment, this digital demeanor—and behind it lies the utopianism of a lot of digital culture—the tools to uncover so much are in our hands; ours and those of people who haven't had access to this kind of cultural tool before.

LHL: Our memories may be gone, but they're certainly recorded now in a way that was not possible before. They're retrievable. What will be preserved and archived will depend on the priorities, cleverness, politics, and rebelliousness of each generation.

MS: Right, you're talking about the will to conserve. It's a task to conserve, to rework, precisely in the way that we took that box of stuff connected to 1972 and reworked it in 2006, 2007. That's the only way the past is going to keep going. It has to be taken up and reworked. So, in the digital proliferation of all this stuff—from the mundane, the quotidian, the everyday of people's lives—we have to see value. The only way it's going to survive to give a new angle on the present, or to be the basis for a new kind of understanding of the everyday history of the twenty-first century, is if people take it up in terms of those energies you've just described.

Michael Shanks and Lynn Hershman Leeson

They've got to want to do it. Material preservation won't work. Information is a verb. You have to take things up and rework them, remix them.

LHL: To make them alive.

MS: To make them live again. It's reincarnation, literally. You incarnate. You give them new material forms that you engage with.

There is all this stuff, so much stuff. I think the great prospect is that some unexpected components of today are going to be taken up and remixed and reworked. Not the great, grand stories of history, not the great accounts proffered by the victors and the great, powerful figures of today. But rather the mundane, the everyday, the stuff that really makes life what it is. That would be fascinating.

And, actually, this is what archaeological science has always offered—accounts of everyday life with which we can all identify and yet find uncanny. It may simply be a thumbprint upon an ancient pot that connects an inconsequential past moment with the present; it may be the evidence of the lives of those who built a place like Stonehenge. It is the archaeological focus on the everyday that many people find fascinating.

LHL: Because these are the relics of ourselves. ∞

Thirteen

Who Makes Science?

Lawrence Krauss and Natalie Jeremijenko

The physicist and the artist discuss science as a public enterprise.

Neither Lawrence Krauss nor Natalie Jeremijenko can be easily categorized. Krauss's deep commitment to public understanding of science advocates for the role of the scientist in the cultural and political landscape. Jeremijenko's scientifically informed, socially conscious art asks who is, and isn't, participating in science. *Seed* invited them to Butter restaurant in New York City to discuss how their views on science as a public enterprise might compare: How do they each define progress? Who do they feel is responsible for science? And what, exactly, is "thinking like a scientist"?

Lawrence Krauss is a theoretical physicist at Arizona State University, where he is Foundation Professor in the School of Earth and Space Exploration, and inaugural director of the university's new Origins Initiative. He has worked with several science museums, received many awards for his research and writing, and appears frequently on radio and television around the world. Krauss is the author of numerous scientific publications as well as popular articles and books, including *Hiding in the Mirror: The Mysterious Allure of Extra Dimensions, from Plato to String Theory and Beyond.* His latest book is *Quantum Man: Richard Feynman's Life in Science* and his new narrated symphony, *Cosmic Reflection*, had its world premiere at the Kennedy Center in Washington, D.C., in November 2009.

Natalie Jeremijenko is an artist whose background includes studies in biochemistry, physics, neuroscience, and precision engineering. Jeremijenko's projects—which explore the dynamic between humans, technology, and nature—have been exhibited by several museums and galleries, including MoMA, the Whitney, and Smithsonian Cooper-Hewitt. A 1999 Rockefeller Fellow, she was named one of the forty most influential designers of 2005 by *I.D.* magazine. Jeremijenko is the director of the xDesign Environmental Health Clinic at New York University and assistant professor at the Steinhardt School of Culture, Education, and Human Development.

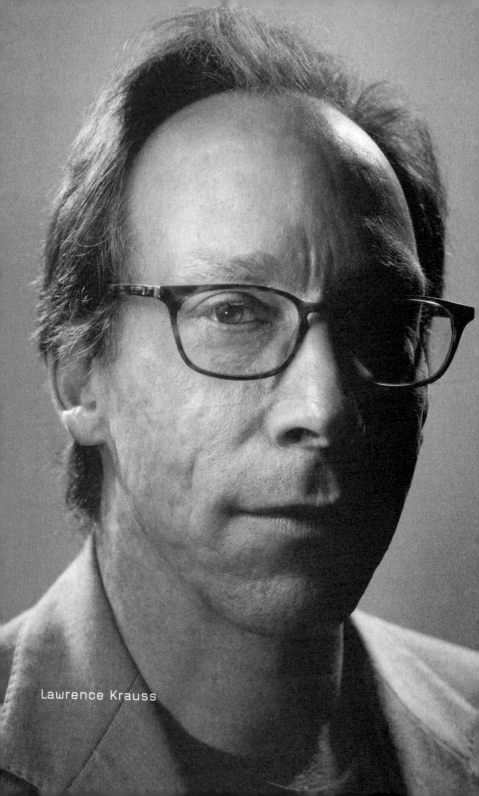
Lawrence Krauss

Natalie Jeremijenko

Natalie Jeremijenko: So, I know you wrote a book called *The Physics of Star Trek*, and I wanted to ask you about it.

Lawrence Krauss: Yeah, sure.

NJ: Because when I came to the States and was working at Xerox PARC there was a language that I didn't speak. Even though I was fluent in Fortran and Pascal and C++, I couldn't speak "Star Trek."

LK: Right—ha.

NJ: I didn't know what these people were talking about.

LK: Yeah.

NJ: When it came to a lingua franca for the entire lab, between computer science and the languages lab and the CSCW lab and so on, the point of reference that everyone shared was, "Oh, like they did in *Star Trek*." So I literally crammed when I got there, borrowed all the DVDs to try and catch up so I would know what the hell they were talking about, what a transporter was.

LK: You didn't see it in Australia?

NJ: Well, I suppose there were Trekkies there.

LK: It's big there. I've discovered that. It's Trekker, not Trekkie though. I had to learn that. Actually I had a similar experience in some sense. And you hit on my motivation for writing the book, which is that a lot of people didn't—and don't—find physics interesting, but they found *Star Trek* fascinating. I've been involved in teaching and writing for a while, and I often tell teachers—though I think it's true for anyone: teachers, car salesmen, artists—that the biggest mistake they make is to assume people are interested in what they have to say. You have to think of a way therefore, as I put it, to "seduce" them into thinking about what you have to say. The *Star Trek* universe was a way to seduce readers into thinking about the real universe, which I happen to think is much more interesting than the *Star Trek* version. But I had no idea myself, at the time, how deeply it was ingrained in the consciousness, not just of Americans, but of Australians now, and Canadians. And I had to cram, too. I was terrified of alienating 20 million *Star Trek* fans, maybe more. The interesting thing about it was I assumed that it was mostly fourteen-year-old boys, but what I discovered was the show was intergenerational, it was gender nonspecific, and, at least in my experience, there were not only doctors, but also lawyers and people from all persuasions who were into it.

NJ: Right, yeah.

LK: And then I started thinking about why and, you know, very often science fiction presents a dystopic view of the future, where science is bad and scary. And one of the things about *Star Trek*—and it's one of the reasons I think it was so popular—is that it's based on this notion, which may or may not be true, that science cannot only make the world a better place, but it can make people more civil and understanding; it can civilize us. And what a weird view of the future, that science can actually do that. I don't know if that vision isn't the most unrealistic thing about the show in some sense. I think science should civilize people. Has it civilized people? That's not clear to me at all. But that's one of the reasons engineers and scientists are so drawn to *Star Trek* I think. Because science, in some way or another, is the hero.

NJ: Well, there's clearly the idea that science and technology can transform society, or do. But popularly, we mistake technological transformation for social transformation. So we recognize social change through concrete technological change. In a movie, we can recognize the year the car was made, or whether they're using cell phones or transporters.

LK: Yes—ha.

NJ: But actually I studied it very hard to figure out what was appealing, because it was this ubiquitous reference, but when I watched it, it seemed like a reasonably fascist, militarized society, right? They go on these colonizing journeys to dominate other planets. The characters were either imprisoned in the *Enterprise* or in some barren landscape, usually under attack, and with no cultural exchange, no trading of knowledge, no fun. Everything was very austere, and nylon. And yet it was intriguing, it had to do with the possibilities; what you could do with scientific knowledge. So that's what I found interesting, that these scenarios developed by science fiction are actually often profoundly socially conservative. Social change doesn't cause, or drive, the technological change. It doesn't even inspire it. But it's there.

LK: Right, yeah, society isn't reimagined in quite the same way as the science, or technology, is.

NJ: But it brings up the question of what the cultural role of scientists is, and whether science can provide a hopeful view of the future, or participate in reimagining society. At PARC they often quoted Alan Kay that the best way to predict the future is to invent it, and we thought we were doing that.

LK: So do you think that scientists need translation devices to function within their own environment?

NJ: Yes, I think so.

LK: To talk to each other? Or also to translate what they're doing to themselves?

NJ: Both, I think, and to translate between disciplines. Joan Fujimura wrote a great book on how the molecular biology department was created at Berkeley. Organic chemistry and biology spoke totally different languages. They had no way to translate between the atomic spins and the biological phenomena that they were dealing with. And her claim was that it was the lab techniques that became the translation device for figuring out what techniques could represent something else, around which they could say, "Well, we would call this this, and you would call that that."

LK: It's interesting. The boundaries have already vanished a lot between physics and chemistry but they're really going to vanish between physics and biology, in the sense that biologists have been

Lawrence Krauss and Natalie Jeremijenko

using the techniques of physics a lot. I remember when I was a graduate student and I got depressed, which I often did when I was a graduate student, and I wanted to quit physics. And I thought of going to do a PhD/MD program in Boston, and I remember speaking to a man who was chair of a biology department at Harvard and he said, no, don't do biophysics, because that's a field that's not of interest to either biologists or physicists. And that was probably true in the late 1970s, early eighties, but it's different now. And the reason, I think, this idea of translation devices is interesting is because we have these disciplinary departments that don't reflect the emerging research, which is cross-disciplinary. It's not just the techniques of physics that are being used in biology, the biological systems themselves are becoming interesting to physicists. Molecular motors are now interesting. Questions are emerging like, Are there new physical systems in biology that you can study as a physicist? Are there new laws that guide how those are either created or function that don't exist in nonbiological systems? And so the techniques are the same, but the questions are new. Still, it is extremely difficult to translate. You have to find techniques to explain to yourself, as a scientist, and to your colleagues what you're doing and why it's interesting. Again, I think it comes back to this idea of seducing.

NJ: You're really into seduction.

LK: Yeah, well, because I think, for better or worse, it describes a lot of what we do in our lives, right?

NJ: That's a very romantic view, I think.

LK: Isn't it the same thing that you've called "strategies of persuasion"?

NJ: Yeah, I suppose, but without the innuendo.

LK: Well, yeah. You're right. But persuasion, if it's effective, is really seduction, right? Because when you're persuading someone, what you're really doing is convincing them to think what you want them to think, but to be happy about it—instead of forcing them.

NJ: Well, you're right. So let's agree, it's a persuasive strategy.

LK: Okay, I'll call it persuasive strategy, but I'll have to get used to that. But I think this happens within a field as well. I remember, I was on a visiting committee at MIT, and these students tend to think they're going to be successful because they're good at what they're doing. But, in fact, a large barometer of their success will be how well they can communicate what they're doing. Not just to the outside public, which most scientists don't necessarily have to do—though I think that's important, too—but within the field, or to your company. It isn't just what you do, it's often how you present it. And, traditionally, we've spent very little time educating our students on how to communicate. So strategies of persuasion, I think, are vitally important within the field. But—and I should be very clear about this—while I understand science as a sociological phenomena, I do believe in objective reality and I do believe that, ultimately, important science wins out in spite of the social constructs and the social or peer pressures to do certain things. So I think that ultimately people realize what's truly significant and recognize it, even if it takes a little bit longer because it's not presented well. I don't think it's all persuasion.

NJ: Professionally, you have to believe that.

LK: Maybe. But no, I think that's what makes science special. As a scientist and someone who tries, for better or worse, to extol the virtues of science in a society that doesn't appreciate many of those virtues, I think that ultimately the good stuff wins out even if it takes a while to do it. Because the final arbiter of success isn't people. In science, it's experiments. It's the ability to make it work. If it works, then people buy into it, whether they like it or not. And I really think that's profoundly important. That, and the oft-misunderstood fact that science doesn't prove things to be true. Science only proves things to be false. That's all it does. But that alone is something that doesn't happen in almost any other area of human activity. The fact

Lawrence Krauss and Natalie Jeremijenko

that you can say, "That's garbage, don't talk about it anymore." The earth isn't flat. We don't need to have critical thinking classes to debate or discuss it. You just go around it, end of story. And the ability to throw out ideas that aren't productive is, to me, what makes science unique and what allows for progress. You don't have to keep wasting your time on the wrong things, because the wrong things are obvious. The right things may not be obvious, but the wrong things should be. And if I could just convince people of that, I think it would go a long way to getting people to have a perspective of science that is useful.

NJ: It's interesting that you would immediately see me as a relativist, which is fine.

LK: Well, I don't know, but I was kind of hoping that I could provoke that.

NJ: Well, I'll string you along, because it's easy to say that science has this experimental evidence and that the messy, real-world, technosocial kind of stuff—political reality, the world we live in, how we actually know whether a technology is good or not—

LK: When you say "good," what do you mean?

NJ: Well, that's just it. How do we know, right?

LK: We know whether it works or not, but whether it's good or not is a totally different question.

NJ: Exactly. It's a messy, hard question that's hard to do an experiment on.

LK: That's why I don't do things that are so messy. The universe just works, whether it's good or not.

NJ: To take scientific knowledge production and methodologies, science itself, and hold them outside of society—the messy, icky world that doesn't apply—is, I think, the big tragedy. And certainly, I'm not going to claim that nothing is real, everything's constructed. But, if we understand our scientific and technical knowledge as constructed within these social constraints, I think we can do better science, better technology, and have better ways to get at what we mean by something being good or being progressive, without saying, "Okay, scientists know how to tell us the answer on this, and everyone else can shut up."

LK: Well, that certainly is wrong. I agree with much of what you said, but there's one thing I guess I disagree with: I think when it comes to technology and society, you're absolutely right, but I'm not convinced it would help scientists do better science. And of course our perspectives on science are different, because your experience is, I don't know whether to use the words "more applied," but it is. And these are emotionally charged words, "fundamental" and "applied." But I think scientists do their best science when they're not concentrating on anything else. I've always been sort of a political animal and worried about science in society, and young people ask me, "How can I get involved and what can I do?" And I usually tell them that if they're any good as scientists, the best thing they can be doing is science, and that as they produce science, the opportunities for them to speak out and to impact on other issues will increase and then they should take advantage of those opportunities. But science, in my mind and from my experience, is best done when you're not thinking about anything else other than the problem you're trying to solve.

NJ: So how do you ultimately answer the question that kids came to you with: "How do I participate?" Because the fact is, when it comes to scientific knowledge production, most people don't know how to participate.

LK: Yeah, exactly. That's true.

NJ: And in a participatory democracy this is, I think, the interesting question: *How do I participate?* How does what I say count in any way? How does my contribution have any value whatsoever? And a consumption-based society, where you're just the dollars you spend, doesn't give you the sense that you have a unique intelligence that you can bring to bear in understanding or contributing to knowledge production, right? And if I would characterize where we are now, with respect to the ways that we have educated and recruited scientists and the way we do science, it is to privilege what are effectively these royal societies and these internal conversations about science.

LK: It's a privilege, absolutely.

NJ: But that's changing. And we're facing an opportunity to structure participation.

LK: Do you think it's possible for everyone to contribute to the generation of knowledge? I don't know if that's realistic.

NJ: I think so, yes.

LK: Well, I mean, as part of experiments, perhaps.

NJ: Well, okay, let's give a few concrete examples. There's globe.gov.

LK: Or SETI@home.

NJ: Or folding@home for protein folding.

LK: But those are largely, I have to say, illusory.

NJ: Exactly. They create this wide, easy venue for contribution where you can donate your spare processor cycles to this larger scientific endeavor. It's like paying a membership to the Sierra Club or joining the American Museum of Natural History. You've done something and you've contributed something, but you haven't contributed anything that has required your particular, personalized input.

LK: It wasn't generated from you.

NJ: Right. And so, the National Science Foundation, the Globe Project, and SETI@home, I think these are not the paradigms of how we can structure participation productively.

LK: Oh, I agree with you.

NJ: So to characterize my own efforts to explore how we can, for example, I started these blogservatories for the One Tree Project.

LK: Yeah, which I was reading about.

NJ: Right, we planted one thousand cloned trees around the San

Francisco Bay Area. I put them in public spaces because now anyone can ask questions and blog on behalf of the trees—each of the trees has its own blog. "I saw a bird in there," "I watered it today," and so on. So because it's outside the controlled context of the lab, people can contribute speculations on why the trees look different, and what's causing that. On the ground, the climatic conditions that are producing such incredible variation in these trees are far too complex for any one person, or any one discipline, to understand. Yet if we draw from the diverse speculations and observations and weird ideas of everyone, I think we have stochastically a much better chance of getting somewhere.

LK: Okay, yeah. I think it's a great thing, but are they really making a vital contribution to the progress of science? Or are they having the experience of learning, of becoming aware of the practice of science? I happen to think science and the arts are the same in many ways in that, at their best, they force us to reexamine our place in the cosmos and reassess our role, where we're coming from. That's more important than the technology in the end, for me. So this project may be forcing people to do that. Realizing that the trees' variation is extremely great, regardless of the fact that they're cloned, or realizing that science is based on lots of individual bits of data and adds up, you don't always know where it's going. All of those things are great.

One of the ways to get people to participate in science is to get them thinking like a scientist. If I can induce people to think like a scientist in one context, then their approach to problems in other contexts may be better. So for me, participating in science is not participating in the progress of research. Which I think, realistically, is restricted to those people who are doing it on a professional basis. Einstein was a patent clerk, but he was also an educated person who was vitally involved in what was going on. Unfortunately, he still created this unrealistic notion that anyone can make a breakthrough. It's increasingly unlikely for that to be the case. But I think what anyone can do is have the experience of science, and therefore be a part of the scientific process, which is vital to our society. And it seems to me that's what you're really giving them—an experience of science.

NJ: Right, because if you asked the National Science Foundation and the Globe Project, "What is thinking like a scientist?" they would tell you in their kind of pedagogical outreach exercises, it's kids going around and collecting the temperature and the precipitation rate. Basically doing cheap labor, cheaper than graduate students.

LK: And it doesn't seem to be creative in any way.

NJ: Right, and kids hate it. It may as well be market research, right? They fill out their little surveys, they hand it in. I would argue that's not thinking like a scientist.

LK: Right. As to what is, I think the key experience—and I wish every single student would have it—is to have some cherished notion that you absolutely believe to be true proved false. That's the experience of science that I think is the most beneficial and most characteristic of the greatest and most important advances in science. It opens your mind tremendously. We assume these realities about the world, and the progress of science is often associated with taking those realities and showing that they're wrong. Something you thought was so wonderful and beautiful and deeply ingrained in your being, chuck it out like yesterday's newspaper. And if you're able to do that, then you're really doing science.

NJ: Right, and that's not at all a part of what the NSF is now funding. Thousands of school kids are now out there "Participating in Real Science," as the tagline says. Yet they're not allowed to ask interesting questions. They're not speculating on why a bird is landing in this tree, and not that tree. They're not figuring out why they're unique. And they're not being questioned. That's a privileged stance when knowledge is organized around experts, and one that undermines the capacity of people to be able to draw on material to form an opinion on their own.

For example, there's not a person who hasn't heard about global warming, right? They've all seen *An Inconvenient Truth*, and moreover, thirty years of environmental activism and work by scientists, activists, lobbyists, and other people have rendered environmental issues global enough to be newsworthy.

LK: Yeah, although people still don't have any perspective on what the real issues of global warming are.

NJ: Exactly. And what I find interesting is that there's now this general, global anxiety without the capacity of translating it into anything local and actionable in any way.

LK: True. Yeah, interesting. Especially a big problem like that—it seems like it almost paralyzes people because there doesn't seem to be anything you can do. And when you talk about buying carbon credits or whatever, it just seems so artificial.

NJ: And what we've missed is the idea that science is not the singular expertise. There are lots of interesting ways of asking questions, drawing on material evidence, finding ways to iterate between things we try out. "Well, this theory is still explaining this, we can try it out again." Or "This theory has collapsed," et cetera. There are ways to build a dialogue with the material world, which is the stance of physicists that I like. There are many stances of physicists that I don't like.

LK: I'd like to hear about those. But I agree, that's a part that I like. You're constantly in dialogue.

NJ: But it comes back to the question of who participates and how. How can individuals, without a master's or PhD in environmental science or boundary layer physics, and without having published a peer-reviewed paper, have access to material evidence, or permission to ask questions and to draw on the facts? That's where, in the cultural world, science has a monopoly on material evidence.

Lawrence Krauss and Natalie Jeremijenko

LK: That's interesting, I haven't thought of it that way. In some sense, you're saying art, to the extent that you call it art—I don't know whether you want to call it that—gives people an opportunity to experience science that the traditional scientific methodologies and sociology may not. Through art, you can get people to participate in what you would say is a scientific experience. Would you say, to some extent, that's what you're trying to do?

NJ: To draw on material forms of evidence, like physicists do. To actually test their ideas on the stuff around them.

LK: So can I summarize it by saying using art to seduce people? ∞

Fourteen

What Is Human?
Will Self and Spencer Wells

The writer and the genetic anthropologist talk about place, identity, and what it means to be human.

Will Self is a walker. In documenting his long-distance journeys, he explores the impact of place on the human condition. Spencer Wells travels to remote areas of the world to genetically reconstruct the migratory patterns of early humans. In late 2007, they made their way to New York City, where *Seed* asked them to consider: Will a global monoculture erase evidence of our diverse histories? What is the evolutionary consequence of urbanization? What, ultimately, does it mean to be human?

Will Self is a writer and satirist who has authored novels, short stories, essays, and newspaper columns. In the acclaimed book *Psychogeography*, a collection of his columns in *The Independent*, Self charted his walks through urban and rural areas, addressing the relationship between place and inner state. Self's published work includes the novel *Great Apes*, which takes place in a world in which chimpanzees are self-aware, and the short story collection *Tough, Tough Toys for Tough, Tough Boys*, for which he won the Paris Review's Aga Khan Prize for Fiction. In his new book, *Psycho Too*, Self teams up again with cartoonist Ralph Steadman for another raucous exploration of environments—both on the outside and within.

Spencer Wells is a geneticist and anthropologist whose work helped to establish Central Asia's critical role in the dispersion of humans around the planet. A National Geographic Explorer-in-Residence, Wells directs the Genographic Project, an effort to collect DNA samples from around the world in order to better understand human migration over the past sixty thousand years. He is the author of *The Journey of Man*, a book and documentary that traces the origin and migration of modern humans using the male Y-chromosome, and *Deep Ancestry*, an overview of genetic anthropology. His third book, published in 2010, is *Pandora's Seed: The Unforeseen Cost of Civilization*. Wells is the Frank H. T. Rhodes visiting professor at Cornell University.

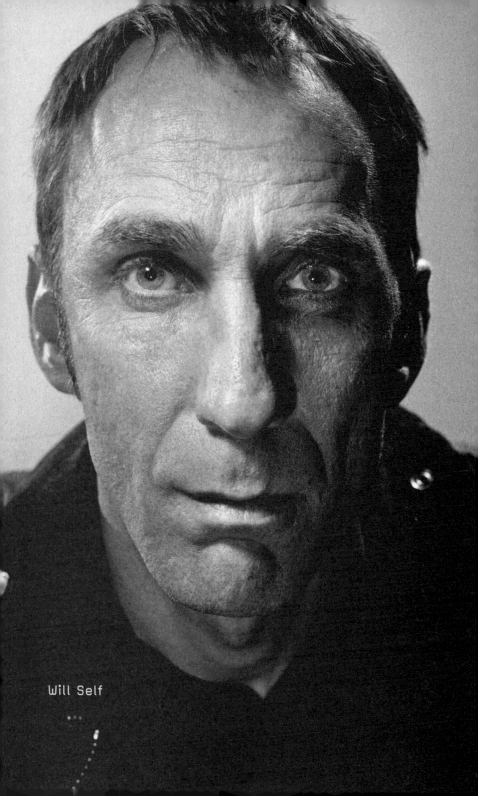

Will Self

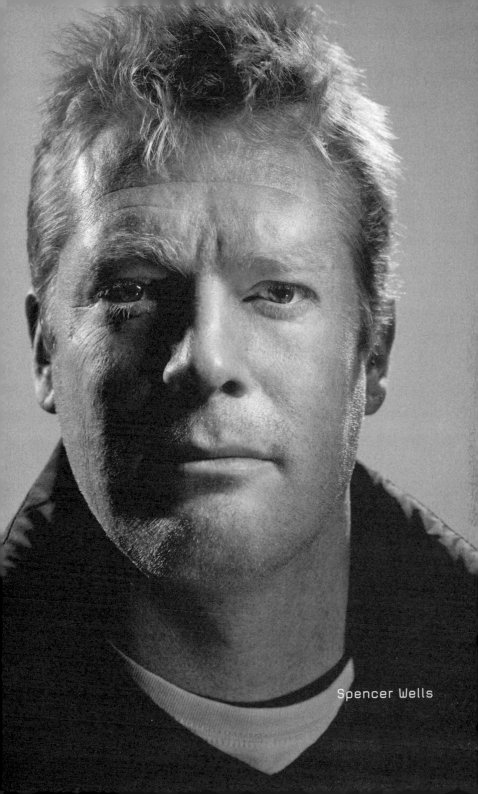
Spencer Wells

Will Self: I once walked from London to Harwich, along the Essex Way.

Spencer Wells: Ah, wow. How many days did it take?

WS: Three and a half days. It was a beautiful long-distance path, the Essex Way, that avoids all urban centers. The bizarre thing about a walk like that—through one of the most densely inhabited countries in the world—is that I didn't meet a soul.

SW: Ha—no kidding.

WS: Just the act of becoming ambulatory cuts you out of contemporary culture.

SW: So you can become more of an observer, I guess?

WS: Completely. Or a Viet Cong insurgent from the past, creeping through the tunnels.

SW: Ha—right, complete with rights of way.

WS: But yeah, I've always walked. My father was an academic who specialized in urban region development, and he was a walker. It's partly a connection with him. He died about ten years ago. And I think it's also bound up with psychogeographic concerns, with mapping these unrecorded places, in a sense.

SW: Your personal *dérive*.

WS: It is a personal *dérive*. Though my *dérive* is quite purposive in a way, unlike the Situationists' *dérive*, which was, of course, random. They "drifted" and wanted people to become aware of their surroundings. But it has some of the same things associated with it—a way of hitting against what I call the man/machine matrix, and against the way globalized travel destabilizes us and alienates us from our own environment.

SW: It does, it does. I mean, you know, because you go on these sorts of book tours often enough that you fly into a city and you could be anywhere.

WS: Yeah. On this swing through North America in particular, having done book tours now for fifteen years, I was particularly concerned to redeem that. So I've been doing these airport walks, walking from airports, or taking long walks in every city I visit. And doing some strange walks, too.

I was very struck by the final chapters of your book where you discuss the way in which globalization is essentially erasing—

SW: The history of our species, yes.

WS: Yes, and I wonder whether or not my impulse to walk were not part of a . . .

SW: Luddite response?

WS: Well, a conservatism, at an unconscious level, to retain that, and to somehow get across to people that that's what we're losing. In my own way I'm perhaps chiming in with that.

SW: I think there's something inherent in humans that, yes, makes us want to migrate, but also to have that connection to place, even though we're moving. I think there is something of a wanderlust in our DNA, something that makes us want to explore a little bit further, but at the same time we want to actually be in the place. The way we travel today, you're not in the place. There's never any "there" there.

WS: Yes, there's no "there" there. I think what I read of yours was one of the clearest and most succinct statements—your view put me in it much more nearly, in a way. Yet, still, I think one of the problems is that it's so hard for us to conceive of things on such a grand scale. You're talking about the wave theory of migration into Europe, which is against the diffusionist theories; you're talking about these approximate measures with which we might conceive of populations being supplanted and moving in.

SW: Mm-hmm.

WS: And speaking as a writer—as somebody who's mostly preoccupied with the zeitgeist—this multigenerational perspective, this perspective of millennia and of great distances, it's all so hard to wrap your head around.

SW: And we toss around millennia like they're nothing— "Approximately ten thousand years, give or take a few thousand." It's many, many lifetimes. So, yeah, I think it is quite difficult to personalize it.

That's why the DNA, I think, is so important. Because if you test your own DNA, you can actually connect back to an individual, a person, who actually lived in a particular place at a particular time. I don't approach it as a genealogist. For me, it's a grand historical quest. I'm using DNA as a way to study history.

WS: Do you think that we are in a race against time here? Do you think there are major questions to be answered?

SW: I do. The diversity of humanity is being subsumed into this global monoculture, if you will. Within the next few generations the genes will still be there, but they will be all mixed up and they will have no context. Socially, I think it's possibly a very good thing. But it just means that we're going to have erased who we are genetically, these patterns of variation that are relatively recent in the grand scheme of things.

WS: So, in going to these remote locations, is that what you are preoccupied with at the moment? Collecting DNA samples?

SW: That's a big part of what we're doing. It's also educating people about all of this, about DNA. Like it or not, the twenty-first century is going to be the century of personalized knowledge of your genome. It's going to be one of the first things your physician looks at when you come in for a diagnosis. That's coming within the next generation, possibly within the next five to ten years.

WS: In your book, you couldn't avoid a note of wistfulness. That's something that I've encountered in other people who write and think deeply about prehistory, a wistfulness about the huntergatherer period, about the late Paleolithic.

SW: Oh, absolutely, yes. Noble savages. That's actually the subject of my next book.

WS: Is it?

SW: Yeah, it's all about the changes that have happened since the development of agriculture and how, even from the very beginning, they seemed to be bad for us. Anyone who has spent time with hunter-gatherers—have you ever spent time with the San Bushmen, or any of these other groups?

WS: No. I've spent a lot of time with heavily deracinated Australian Aboriginals, but there's enough of it there.

SW: You see this amazing natural historical knowledge they have of the place, and this intense connection. You sit around in the evening after they've been out hunting all day and you listen to them tell stories. You can't understand a word they're saying, but you still have this sense of something amazing being transmitted between them, and they seem genuinely happy, and yet they have nothing. I think that's an important lesson.

WS: It is a form of natural magic, of course, for Westerners. Lévi-Strauss said that the thing about magic was that it works.

SW: Right. It would have been myth otherwise.

WS: One of the other things that grabbed me, as it always does when I read about deep time, is the residual in our mythology of "nonhumans," perhaps of hominid species that we lived alongside. Think of the enormous grip that *The Lord of the Rings* has on us, as a story cycle.

SW: Yes.

WS: I think that's because the folk memory of that period in prehistory, which must have been a long time, is still with us. It's why fairies and elves and so forth persist.

SW: And possibly the yetis in the central Asian mountains. I think that they, realistically, could have been a surviving band of Neanderthals that people encountered until the past few thousand years.

WS: All of that mythology is clearly about that, isn't it?

SW: Yeah.

WS: We have a mutual friend, Chris Stringer at the Natural History Museum in London.

And he recently showed me a hand ax that dates to four hundred thousand years ago, dug up in England. So who made that hand ax?

SW: Probably a species we would call *Homo heidelbergensis*, which we believe was the ancestor of Neanderthals. So it's a hominid cousin but not a direct ancestor.

WS: This hand ax is a beautiful piece of work to me in terms of its adaptation. There are some theorists of the Paleolithic who view these other hominid species as being more technically expert than we perhaps might imagine. I wonder about their demise in that sense—and I suppose it wouldn't be called genocide, but perhaps specicide, in this context.

SW: Call it genocide if you want. But I don't think it was something that we set out to do, in the case of the Neanderthals. It's just that they were perhaps better adapted to forested countryside, which is what Europe was until the worst part of the last Ice Age when the grasslands really came in and the tundra moved south. We were better adapted to living in open country and we had larger group sizes and better hunting techniques that had probably been developed on the steppes of Central Asia, among other places. We just out-competed them.

WS: There was a strong inclination toward speciation among humans, but we now have to reassert our connection with nature and the environment at the end of millennia and millennia of doing exactly the opposite. Those Neanderthals have nothing to do with us.

SW: Well, it's a generalized xenophobia, to recognize things that are like us, and are, therefore, to be trusted. It's scary to think that might have been something that we've been adapting to do, for hundreds of thousands of years, and now suddenly, it's not a good thing.

How do you get past that in this modern world?

WS: But it is a paradox, isn't it, if we're going to be attracted to the other to the extent that it's no longer the other? With the level of miscegenation that we have at the moment, the other isn't going to look like the other anymore.

SW: Eventually, we will all look much more like Tiger Woods, perhaps.

WS: Yeah, then these key markers of who we once were will be eradicated.

SW: Then we'll develop new ones.

WS: It seems to me that humanness has been a question of refinement.

SW: That's the way evolution works. You don't typically have a revolution, you have slow changes, and evolution tinkers with what you have available.

WS: So what is to say that we're any more human now than we were before the great leap out of Africa? Before the agricultural revolutions of the Neolithic period? What is it that makes us any more human?

SW: I don't know that we are more human than we were before we left Africa. This whole Neolithic transition is something that really bothers me—the things that were set in motion simply by growing a larger population.

Suddenly you need a hierarchy to control the people and you need organized religion, and those have had a bad long-term effect. But no, I don't think we are any more human than we were say one hundred thousand or seventy thousand years ago when there's evidence of this change in behavior that made us fully modern.

WS: Could we be less human?

SW: Yeah, I think in many ways we probably are less human than we were as hunter-gatherers.

WS: I'm toying with these notions because it seems to me that this sadness, this inability to reconnect with the natural world, may be fraught with our understanding of what it is to be human.

SW: We've probably changed more since the dawn of the Neolithic than we did in the hundreds of thousands of years leading up to that. Basically what we're doing is adapting to the culture we created, which is a frightening thing because the culture, in a sense, has become a living organism of its own. It's almost like a virus the way it's taken over. The greatest adaptation seems to have come from the change in diet and the change, perhaps, in shelter and making clothes, and all these things that happened as a result of the Neolithic.

WS: And is it encrypted in our genome that we'll be fatter, for example?

SW: Well, if you believe in something called the thrifty genotype, which was suggested back in the 1960s by an American geneticist and physician called Jim Neel, when he was studying groups in the Amazon and other places. He looked at the transition as hunter-gatherers or subsistence farmers moved into the twentieth century and took on all the attributes of mass culture. And often people do become incredibly obese. You see this in groups like the Pima Indians in the United States, where something like 50 percent of the population is diabetic.

WS: Do you look around at airports and think, "I can almost see genes mutating in there?"

SW: I don't think we're changing fast enough at the genetic level. It's interesting because now, suddenly, we have the technology to choose the direction we want to take. But now I think the shape of humanity, the morphology, is in the process of changing. And I think we'll go through a process of adaptation and then we'll probably fast-forward that by changing our own DNA.

WS: Can that be construed as an evolutionary change in itself?

SW: It can, yeah. We were kind of preadapted to do this, if you will. By becoming clever enough to develop the technology over the last few centuries, it's almost predictable that we would be able to do this at some point.

WS: But it's not going to be true of dwellers in the favelas of Rio and São Paulo?

Will Self and Spencer Wells

sw: No, so I think there's going to be a divergence in the world.

ws: It sounds H. G. Wellsian to me. It sounds like time travel or a picture of the Morlocks living underground, all hairy, and—was it the Eloi?

sw: The Eloi, yes—rich, thin, attractive. It's entirely possible. We now suddenly have the power to change things many generations down the road. When you're choosing the genes to put in your children, you're choosing the genes that go into the grandchildren and the great-grandchildren and so on. How do you know that the genes you're choosing are going to be good in ten thousand years or one hundred thousand years?

ws: You're worried, but you're a meliorist; you believe that it's possible that these problems can be dealt with. Do you think it's a scientific attribute to be melioristic?

sw: I think it is—to imagine that we're clever enough to figure out an answer to some of these problems, to find a way around them, to find a fix.

ws: Yeah, definitely the artistic temperament is can't-do, and don't-even-particularly-care—ha! Unfortunately.

sw: I sometimes wish scientists had a little bit more of that attitude. Scientists can be a little too rah-rah about technology.

ws: So in a sense you're saying: We can't see a problem without thinking how to solve it, but maybe that's part of the problem.

sw: Yeah. It gives us license to do anything we want because we know we can fix it and deal with the consequences later. And that may not be the case in every situation.

sw: You've always lived in London, your whole life?

ws: Yeah.

sw: Were you ever tempted to move somewhere else for a while? Be an expat?

ws: Not really.

What Is Human?

SW: Why not?

WS: Well, because it feeds back into some of your work. I think that if you are preoccupied, as a writer must be, by the power of the reflective self-communicating liquidity of felt experience, then language, what you can do with language and its capabilities, is absolutely preeminent and essential for you.

You have the great example of Joyce being a furious expatriate and also an incredible linguist and polyglot. Yet his greatest work is saturated in one dialect and one place, one time and one day. You can't get more localized than that. And it's completely based around walking. I don't think anybody uses a wheeled form of transport. It's a hunter-gatherer knowledge.

SW: Wonderful description.

WS: You have to be a localist in my view to really pull off that obsession. You can write about other things, but there are writers of place and I am one. It's why I've always wanted to stay, and be saturated in that linguistic mulch. Whereas, of course, you move around.

SW: Yeah, incessantly. People ask me where I live and I typically say a series of random hotel rooms and yurts and tents scattered around the world.

WS: Do you think we've evolved to the point where people think they are who they are because they talk a lot on their cell phone and they play with their BlackBerry a lot and they're constantly in flight from one major city to another and from one hotel room to another?

SW: Yeah, modern life is about being disconnected. We rarely connect in a very real way, either to a place or to each other.

WS: Do you think that would have already impacted the genome? Given the levels of chromosome mutation that we tend to look for in human morphological development, how long would you expect it to be before you can put a strand of DNA under a microscope and say, "That is a BlackBerry mutation"?

SW: Ha! A BlackBerry mutation. Wow.

WS: That's a BlackBerry one!

SW: I think there's preexisting genetic variance out there. It's a question of applying the right sort of selection. That's the nature of evolution—things arise and if there is no need for them, if they are not selected for, then they can disappear very easily.

WS: Why would nature have been so profligate in setting up these potentialities?

SW: Well, it's just the background of random mutations. They are occurring all the time, whether we like it or not.

WS: Why aren't there as many for other animals as there are for us?

SW: There are. There's actually more variation in chimps and gorillas and orangutans; between four and ten times as much at the DNA level, because we nearly went extinct about seventy thousand years ago. We dropped down to two thousand people.

WS: I had no idea.

SW: Yeah. We went through a bottleneck. We came back from that, and probably what allowed us to survive, in part, was that change in culture.

WS: I really had no idea about that; it's not well publicized.

SW: Well, it's something I talk about, and Jared Diamond has mentioned it, although not the extreme figures. That comes out of genetics really.

WS: We've now got the two thousand figure.

SW: Yeah, roughly two thousand.

WS: What do you think the population might have been before then?

SW: Probably not much more, maybe between ten thousand and one hundred thousand, but certainly not in the millions or billions.

WS: I think one of the things that inspired me to write *Great Apes* was the imminent extinction of the chimpanzee in the wild, which I think will be one of the most philosophically queasy moments. But I don't think people have reckoned on it at all.

SW: Any extinction, but particularly chimpanzees.

WS: Particularly the chimp, surely.

SW: It's the finality of it and the notion that "these are our cousins, and we're the ones who caused their demise."

WS: Isn't it also like kicking out the ladder beneath us? The connection is then gone between us and the rest of the natural world in a really profound way.

SW: Yes, but we've done that before—we did it with Neanderthals.

WS: Yes, we've done it before.

SW: We seem to have no qualms about doing things like that. We're very good destroyers, as well as creators.

WS: I think, I can't remember the figure, but there are something like only 150,000 chimpanzees left in the wild?

SW: And fewer orangutans.

WS: And fewer mountain gorillas. But if they were allowed to get on with it now, they would be fine.

SW: If we set aside the territory, yeah. It's really that simple. It's just like hunter-gatherer human populations. All they need is to be left alone with enough territory and they will be fine.

SW: Are you optimistic about the future? It sounds like no.

WS: I believe in the Tao. I'm not optimistic or pessimistic. What is, is, and what will be, will be. And indeed, I would argue—the melioristic side of me would say—that our ability to cope with the possibilities will only benefit from equanimity in the face of it.

In other words, what's more likely to power us toward an overheated destruction is a frenzied technical solution. In fact, "Don't just do something, sit there," is probably a reasonable response to the situation. Or rather, "Don't just do something, walk around a bit."

SW: Right.

WS: It may not be like the Lou Reed line in "Beginning of a Great Adventure," where he says something like, "I'm going to breed up a mutant piglet army in the woods." But it's not far off that. And the only way I can see to impact that mind-set is to walk. Maybe it is in the migration, and specifically in the migration that is ambulatory. This may be very old evolutionarily—it may be a mind-set from the past—but we need a physical analog on which to operate. Because presumably we've been chipping rocks far, far longer than we've been dealing with the abstract or the idea of digitized knowledge in that way.

I think the only thing I can do is to try to persuade people to walk. I think that impacts everything we've been discussing. Once you walk, it starts to fall into place.

SW: A Zen-like state.

WS: And a Zen-like state of absorption into physical geography. Because if you are solely concerned with orientation and movement, then the so-called higher faculties don't have a lot to do. There isn't a lot of room. You're not tormented by what the Germans call the earworm gnawing away at you, or the resentment you had toward the guy at the party in 1985 who spilled the drink on you. That goes after a few miles. And I think then we're probably back in the hunter-gatherer mind-set. What I'm saying is, you can be a hunter-gatherer. You can be a hunter-gatherer now. Anybody here in Manhattan, or in any major city, can make that choice to be a hunter-gatherer. ∞

Fifteen

Fractal Architecture
Benoit Mandelbrot and Paola Antonelli
The curator and the mathematician discuss fractals, architecture, and the death of Euclid.

While studying architecture at the Politecnico in Milan in the 1970s, Paola Antonelli was inspired by Benoit Mandelbrot's geometric ideas and visualizations, and eventually wrote a thesis titled, "Fractal Architecture." The two met for the first time when Antonelli invited Mandelbrot to a *Seed*/MoMA Salon, a monthly gathering of scientists, designers, and architects. Just before Antonelli's Design and the Elastic Mind exhibit opened at MoMA, they reconnected to discuss fractals, architecture, and the death of Euclid.

The father of "fractal" geometry, mathematician Benoit Mandelbrot coined the term in the 1970s to describe his investigation into the phenomenon of self-similarity. His ideas revolutionized mathematics and have had a profound impact on several fields across the physical and social sciences. Mandelbrot's numerous papers and books include *The Fractal Geometry of Nature*, which was recognized by *American Scientist* as one of the most influential science books of the twentieth century. In January 2006, he was appointed Officer of the French Legion of Honor. He is currently Sterling Professor Emeritus at Yale.

Paola Antonelli is senior curator of Architecture and Design at the Museum of Modern Art, where, beginning with her highly acclaimed debut exhibition, Mutant Materials and Contemporary Design, and more recently with Design and the Elastic Mind, she has consistently challenged traditional definitions of design. She has curated exhibits worldwide, contributed to numerous publications, and lectures frequently at global gatherings such as the World Economic Forum. In 2006, Antonelli earned the Smithsonian's prestigious Design Mind Award, and in 2007 was named by *Time* as one of the twenty-five most incisive design visionaries.

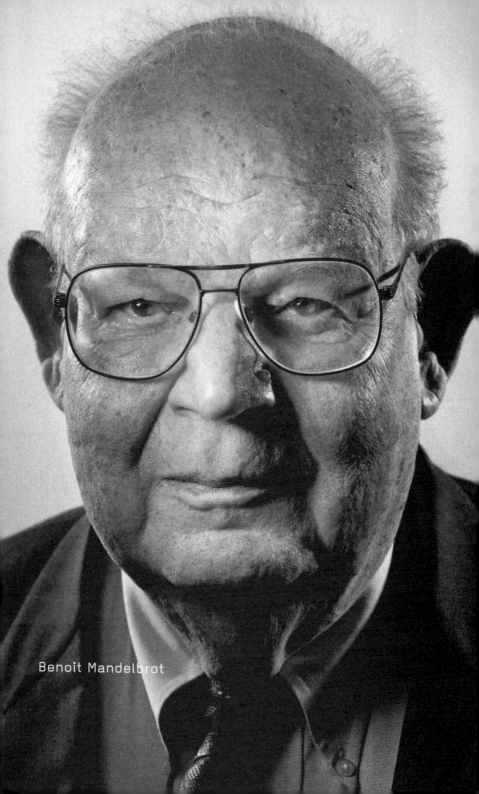
Benoit Mandelbrot

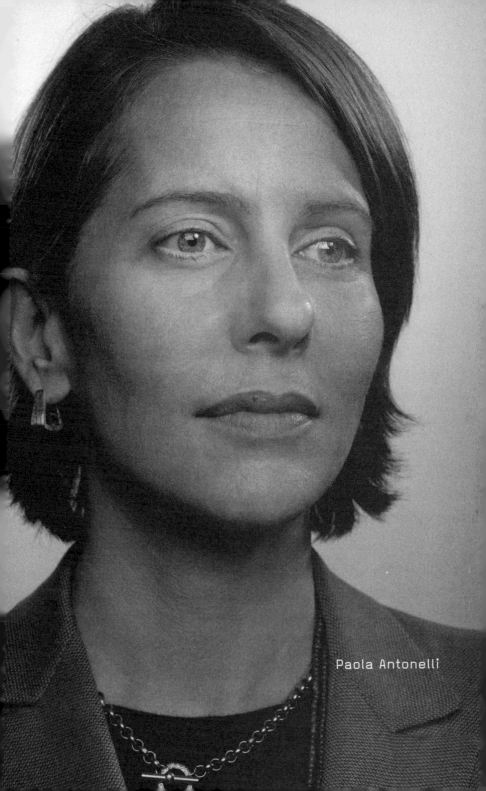

Paola Antonelli

Paola Antonelli: So, here we are. It's eighteen years after my thesis, and I finally get to meet you.

Benoit Mandelbrot: Well, everything happens if you live long enough!

PA: That's right! I'll tell you briefly what it was about, because I just want to have your reaction. I was very, how should I put it . . . very naïve about the mathematics involved in your thinking.

BM: I am naïve about the art.

PA: Well, I hope so. It was not art, it was architecture. In any case, I had tried to read your first book about fractal geometry. Of course, I was skimming through it, not understanding any of the equations, but I noticed something: Some of the most recent architecture—and in particular I was studying the work of Coop Himmelb(l)au, an architectural group from Austria—could not be represented anymore through plans, sections, and elevation. There was no way. Not even with axonometry. Or perspective. Normal geometry just did not work.

Also, you couldn't photograph it; pictures wouldn't render the spaces at all. The only way was to experience them. And somehow, without really having any mathematical or any theoretical proof, I thought there was a connection between your book and this kind of architecture.

So I decided to explore this theory and do my thesis, which was called "Fractal Architecture," on it. Now, thank God, it's only in Italian, so, you will never . . . Oh, my God, maybe you speak Italian!

BM: No, but I can read Italian.

PA: Oh, no, no. But, it's really just very interesting to see how your theories, your geometry, and your work have had tremendous impact on the world, even on people who didn't know about them. When I interviewed Wolf Prix, who is the principal of Coop Himmelb(l)au, I asked him, "Do you know anything about fractal geometry?" He said no.

I just found it really interesting. There's a real impact that your science has on the world, and vice versa.

BM: Well, I've had a life, how to say it, full of adventures, though not always by choice. Things were very complicated during World War II. Altogether it never quite left me the leisure to decide who I was.

PA: Where were you?

BM: I was in central France, in an area you could describe as French Appalachia. There were deep valleys, and people considered Parisians foreigners. It's a very interesting place.

It was not occupied, but it was very closely supervised by the Germans. And one didn't think of the future particularly; the future was so distant. First survive and then there'll be a future. So I never really decided in which field I was going to spend my life—a situation with pluses and many minuses. It's one of the most peculiar and striking aspects of my scientific life.

But over the years I've recognized things that are very close to my work and could not have conceivably been associated with mathematics.

PA: The power of fractals is that they're so instinctive. They're immediately graspable even without knowing there's a geometric law behind them.

BM: Well, that's the astonishing thing—and to me it was an amazing surprise. I was very visual, of course, but I did not view myself as a future scientist.

Mathematics in high school was easy but much less exciting than French history or language. I did well, but it was not something very important to me. Then I stopped school for a while, which turned out to be very important. I went on studying, but my way.

Once back at school, for each problem the professor posed, I had an instant solution—never the same as his. My solutions involved shapes. So I was taking these very dry questions that he asked, and without being particularly conscious of my thinking process, solving them all—near instantly—in terms of real shapes. This took no effort whatsoever. I had, how to describe it? A very freakish gift. In every mathematical question that was asked, I just saw something real that had the same properties.

PA: The things that you saw, were they coming from the real world? Were they coming from intuitions? What would you connect them to?

BM: They were coming from everywhere. During the period I wasn't in school, and couldn't study systematically, I read a lot, whenever I could. So I would remember many things through, how to say, mathematical simplifications.

PA: So you had the intuition and then you would recognize this intuition in the things that you saw.

BM: Absolutely.

PA: I'm sure you know that your work has had tremendous impact in architecture and in design, not only formally, but also philosophically. The idea of the algorithm, of the growth of structures, and the growth of objects. Who was the first architect or designer that contacted you and wanted to talk about it and wanted to learn directly from you?

BM: Well, actually, I think that it wasn't that they came toward me. I came toward them.

PA: Really? Interesting. So, who did you refer to?

BM: Well, a paper I wrote, and that was widely quoted, concerned fractals and architecture. It was in a certain sense a critique of the Bauhaus. A very short paper, but very influential.

I focused on Mies van der Rohe and the Seagram Building because of my anger against Mies van der Rohe's misunderstanding of something I very much care about. By contrast, take Charles Garnier, who primarily designed the opera houses in Paris and Monte Carlo.

He was not very popular, but represented—at least for somebody

Benoit Mandelbrot and Paola Antonelli

with a French education—the kind of principle of what architecture should do.

PA: Meaning?

BM: Meaning, for example, walking toward the Garnier opera house in Paris, from far away, the most striking thing is the roof. You come closer, other things appear, but they are always of approximately the same degree of complication.

Whereas Mies van der Rohe seen from a distance is just a big box. As you get closer you see a grid of windows on the box, and as you get really close, you can see some things of whoever lives behind the windows.

The building itself had the smallest number of scales imaginable. It is very simple to describe. And the architect was proud of it.

PA: Of course he was! He simply was not going after the same effect you're talking about, which is organicism in architecture. That's truly what you are praising. But somehow you also need to have complete abstraction and the simplification of details in order to be able to appreciate organicism. Modern architecture had a reason to exist.

BM: Well, modern architecture had two reasons to exist. One is the desire, on the part of architects, to be different. And the other is the desire, on the part of the builders, to be cheap. Look at modern architecture in early manifestations, for example, in Russian building designs shortly after the revolution—many of which were never actually built, for lack of funding. They were very conscious of the fact that this was not something beautiful. So Garnier, who, again, was not a creative genius, but was a representative of a certain school of architecture, put it very, very strongly. From a distance, you could see something, and as you come closer, you see something else but always of the same kind.

PA: That's like medieval architecture. It's like the cathedral of Milan. Yeah, I understand.

BM: Absolutely, and this is so much more interesting architecturally and aesthetically.

PA: What is really amazing to me right now is how contemporary architects are using the idea that is behind fractals, the idea of a rule that lets them work at different scales indifferently, at least until the moment when the real design application, the reality of the client or manufacturer wanting a building or a toaster, sets in.

I am thinking, for instance, of Ben Aranda and Chris Lasch, who you may remember spoke right after you when we had the salon at MoMA. They are two architects who have founded their practice on understanding algorithms and finding ways to take scientific concepts and translate them for architecture's benefit and evolution.

So it seems to me that it is not only and simply about the formal beauty of fractals, it is the idea of growth that your theory has really given to architects and designers.

And now we're seeing the algorithm become the principle, and the subject of research, for so many architects today. They're hoping that they can ultimately input an algorithm, give it a push, and then all of a sudden an object, a building, a city, and a world will grow out of it.

BM: Well, that would be very exciting and I am very pleased to hear you say it. I have, of course, a good inkling of it. I can speak of other great masters, or unknown masters, who proposed no principle, recorded neither reasons nor comments, but did work along these lines. So the long time it took for this to be codified is astonishing. It's astonishing that the motivation behind these other great works was not more actively pursued. Because they are manifestations of the fact that certain numeric ideas are permanent.

I am not only a scientist, and I find it very important that great architects have very often followed the same path as scientists. And now, it seems, the evolution of these ideas continues in the kind of architecture you describe, this time with a scientific spine.

Benoit Mandelbrot and Paola Antonelli

But in the past, nobody could understand them, nobody could appreciate what was behind them and so they weren't often recorded. But, well, history has its own funny ways.

PA: In a way it is almost a fractal attitude, an indifference to time; the past and the present and the future have the same instinctive approach to things.

BM: Ha, yes. It's mind-boggling.

PA: I would like to ask you about some major phenomena that have happened in the world since the publication of your books, phenomena that seem almost manifestations of fractals in the world. One is the Internet, for instance.

BM: I was well placed to know about the Internet, since of course it became very important when I worked for many years at IBM. And colleagues mentioned to me some strange things about the way in which the Internet became organized. There was no single overall architect and many things were happening by local decisions. A terrible mess ensued and the question was, Can you see any order in that mess? I was pleased to discover some order, though it was not my field.

PA: And what about contemporary architecture? Have you seen the idea of fractals translated in a particularly powerful way in recent architecture?

BM: My influence may or may not have helped, but certainly the mood is different. Most of modern architecture was, how to say, cheap. For example, driving from Charles de Gaulle Airport into Paris, you go by many buildings that are absolutely abominable.

They are cubes of the worst kind and I would hate to live there. Admiration for this simplified art, this Euclidean architecture, which sticks to cylinders or cubes or parallelepipeds, was very short-lived. And most people didn't like it. The profession, I'm sure, had no choice at the time. A few people enjoyed it, a few people got a good name for it. But at this point, I think it's safe to say the idea that perfection is a cube is over.

PA: Hmm.

BM: I remember, afterward, when suddenly fractals became all the rage, at least in some schools. And I was afraid it would just die off like many little fads. But in fact, it continues.

This has been for me an extraordinary pleasure because it means a certain misuse of Euclid is dead. Now, of course, I think that Euclid is marvelous; he produced one of the masterpieces of the human mind. But it was not meant to be used as a textbook by millions of students century after century. It was meant for a very small community of mathematicians who were describing their works to one another. It's a very complicated, very interesting book, which I admire greatly. But to force beginners into a mathematics in this particular style was a decision taken by teachers and forced upon society. I don't feel that Euclid is the way to start learning mathematics. Learning mathematics should begin by learning the geometry of mountains, of humans. In a certain sense, the geometry of . . . well, of Mother Nature, and also of buildings, of great architecture.

Now, do you think I'm just having dreams of grandeur in my old age, or is it true that I provided mathematics with a wider audience? I get letters all the time from high school students, from all kinds of places, and they often begin by saying: "Well, we just realized that you are still alive! We thought you'd be long dead." Which is a bit . . . well, I mean . . .

PA: Flattering!

BM: I'm getting used to it. But what do you think; don't you think that mathematics like this is more alive, warmer? That it is catching?

PA: What really helped fractal geometry and its application in school, I think, is the computer. Having computers in classrooms has been a blessing for all sorts of more visual and more organically based forms of geometry and mathematics.

It helped popularize the idea of fractal geometry and make it become more comfortable and easier for people to accept. And then it also became something for the more elite culture of architects and designers to adopt. I wonder whether the idea of the use of algorithms in architecture was introduced not only by biology, but also a lot by fractals. And the fact that there is more and more science in many architects' and designers' work is very telling. Before, science was kept at bay, and architecture found its inspiration elsewhere; now, science instead appears to be more immediately useful and present in their vocabulary, perhaps because it has gotten so much closer to a real description of the world.

BM: At one point in history a copy of Euclid was shipped from Spain to Italy, and translated. It provoked an extraordinary change in very many aspects of life.

To begin with, it was read by architects and painters; Giotto, a great painter, had no idea of perspective, so he was incapable of representing the beams in that amazing long refectory that he painted.

However, after Euclid became known, his geometry could be taught to anybody. Therefore, there was a moment in history when a mathematics, very different from anything that existed before, came back to the West by the intermediary of Italian painters. This may or may not have contributed to the greatness of the Italian Renaissance.

So mathematics can have a direct influence on everybody's world. Earlier mathematics had developed very separately from the world. Early on, Euclid was very far from everyday reality; but then the world changed, and mathematics became indispensable.

BM: So I know that you are preparing an upcoming exhibit at the museum. I've visited it many times, and each time, it's bigger!

PA: It is definitely bigger.

BM: And each time it's more varied. But tell me, what viewpoint or theory or approach do you hope to foster with this exhibit?

PA: My specialty, my passion, is contemporary design. I'm trained in architecture, and I am proud to spontaneously spot traces of the indifference to scale that you preach; I view architecture, urban planning, design, objects as theoretically the same.

This particular exhibition, which is called Design and the Elastic Mind, comes fourteen years after I started at MoMA. With every exhibition, I've tried to show people the importance of design, and this time I found a very strong alignment with science. Interestingly, both design and science are trying to change their position in people's perception. Science is trying to be perceived as more part of the real world and less lofty than before, and designers are tired of being considered decorators, because they have much more structural roles in shaping people's lives. They really anticipate behaviors and guide change.

Designers take scientific revolutions and they make them usable and exploitable, comprehensible to the average human being. The Internet is an example: it used to be lines of code, and then the designers came. It became an interface, and now we're using it.

So this particular show is about how designers and scientists work together—how they worked together two years ago and how they'll work together in two years. It's about the present. It's about the discoveries that are being made right now.

BM: Ah, interesting.

PA: There's a very strong component of nanophysics and nanotechnology and how they can help shape a model of collaboration for science and design in the future.

Something that is truly interesting, that Peter Galison at Harvard first talked about, is the idea of "nanofacture"—the idea that scientists are compelled to become designers because of the possibility of building things atom by atom. And you might have given them a hint already, with fractals, because it was already something playful that they could do.

Scientists are designing. And designers are trying to learn about science and collaborating with scientists. And together, they are trying to help people cope with the tremendous changes in everyday life—in scale, in resolutions of screen, in contact with big crowds of people.

And what I hope, as I do with every show, is that people will recognize themselves in it. I hope that people will immediately say, "Oh my, this happens to me, too," and therefore understand the role of design—and this time, also of science—in their everyday lives.

BM: Well, it is very encouraging for me, because I'm an old man and, as I always mention at some point, I never made up my mind who I really was, which allowed me to spend my life on many things. So what you're telling me is that I can just relax, because I won't have to decide!

PA: I don't know. You're very responsible for what goes on right now. I don't think you can relax any time soon!

BM: Well, yes, but at least I won't have to become a specialist, because everybody is going to become a generalist.

PA: Generalism is very important. The interesting aspect of your theory is that it was very easy to generalize. And I'm not saying it as a disadvantage; I'm saying it as a quality. It's possible to grasp it, even if you are not a scientist or really versed in mathematics.

So I think that your ideas and your approach were almost the beginning of generalism. And designers are big generalists, and scientists are trying to become a little more generalist because sometimes they feel that they have become too specialized.

But I think you can't sit down and relax quite yet, because you see what happens when architects like Ben and Chris get a hold of you. Discussions go into the wee hours of the morning. I think that the immediate application of your ideas, in design and architecture, has only just now begun to happen.

BM: Great news.

PA: Yes, it is. ∞

Sixteen

Morality

Marc Hauser and Errol Morris

The evolutionary psychologist and the documentary filmmaker debate game theory, Stanley Milgram, and whether science can make us better people.

Errol Morris has made a career of trafficking in moral ambiguity and complexity. Marc Hauser has pioneered research into the idea of a universal morality grounded in biology; Hauser believes humans possess a moral grammar. Morris isn't so sure. The two met when Morris asked Hauser to be part of his short film for the 2007 Oscars. They kept in touch, exchanged ideas, and Hauser attended an early screening of *Standard Operating Procedure*. When *Seed* invited them to Boston, they discovered good, evil, and the science morality.

Marc Hauser is an evolutionary psychologist and biologist whose research is aimed at better understanding the processes and consequences of cognitive evolution. His theory of "humaniqueness" differentiates human and nonhuman animal cognition. Hauser is the author of several books, most recently *Moral Minds: How Nature Designed Our Universal Sense of Right and Wrong*. The recipient of several awards, including a Guggenheim Fellowship and a Collège de France Science Medal, Hauser currently co-directs Harvard's Mind, Brain, and Behavior Program and is director of the Cognitive Evolution Lab.

Documentary filmmaker Errol Morris's critically acclaimed works include *Gates of Heaven*, which Roger Ebert called "one of the ten best films of all time," the Oscar-winning *The Fog of War*, and *Standard Operating Procedure*, a 2008 film about the abuse scandal at Iraq's Abu Ghraib prison. Morris and author Philip Gourevitch co-wrote a companion book of the same name. In 1999 the Sundance Film Festival paid special tribute to Morris, and the Museum of Modern Art mounted a full retrospective of his work. He has received multiple fellowships from the National Endowment for the Arts, as well as Guggenheim and MacArthur Fellowships.

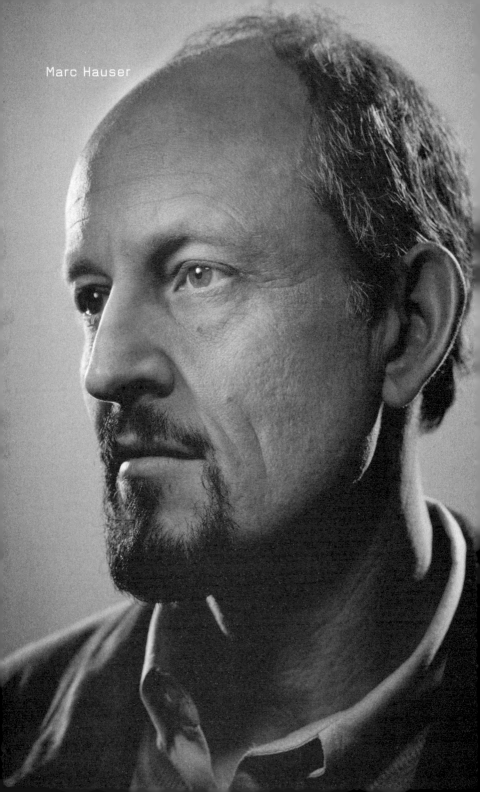

Marc Hauser

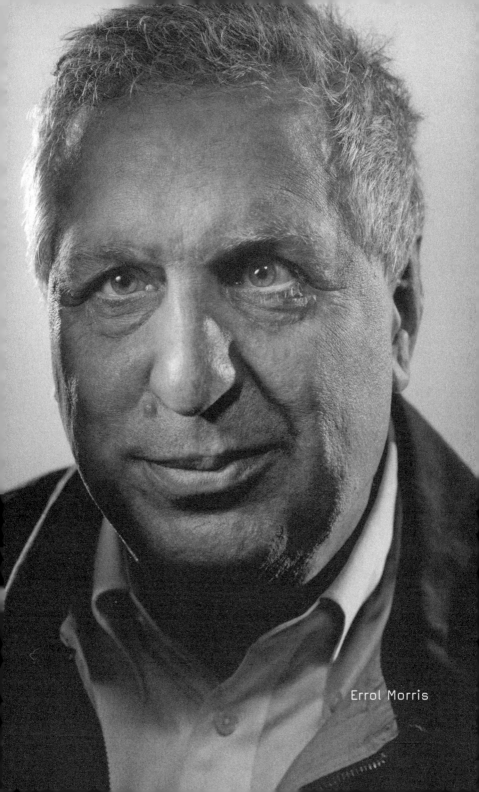

Errol Morris

Errol Morris: I'm not sure that I have any real grasp on morality at all, much less some universal idea of morality. I've thought a lot about what happened at Abu Ghraib, and maybe this shows just a fundamental deficiency on my part, but I've come away even more confused than when I started and more convinced that social science really hasn't grappled with these issues in a way that I find satisfactory.

It's fascinating that whenever we come up against something that is really complex, there is this very deep human need to find a simple explanation that can account for it. If it's something that's really bad, really wrong, people feel uneasy and want to figure out how to distance themselves from it; to tell themselves, "This doesn't concern me. This isn't about me. This is about somebody else, or some other group I don't belong to."

Marc Hauser: Okay, but you left out a huge question: How do you know it's wrong?

EM: Yeah.

MH: How do you even know? How does the human mind know that something is wrong? And once it does, what does it do with that information? Those are deep questions. I mean, they've occupied people for centuries.

EM: Right. One of the stories that I tell in *S.O.P.* is about a suspected insurgent who's brought in by Navy SEALs, interrogated by the CIA, enters Abu Ghraib under his own power, leaves as a corpse on a gurney. So, who is responsible? Who should be blamed? And if they haven't been blamed and are responsible, why haven't they been blamed?

MH: Right.

EM: All of that really interests me.

MH: Well, so what science is doing is trying to distinguish two aspects of the question of how we think about the world. Because even this kind of violence, depending on whose side you're on, will be evaluated differently, right?

EM: Yes.

MH: So, when the Nazis got together to exterminate the Jews, from their perspective, wanton killing of Jews was not wrong. It was perfectly right because Jews were "the other." You map a distinction by recruiting the most powerful and violent emotions you can—disgust, hate. You call the other parasitic vermin to recruit the most incredible imagery. Once you do that, the emotions wreak havoc and you feel perfectly justified exterminating the other.

So this is where I think some of the universality comes in. Say I tell a story about a violent episode but I don't say who's involved. I think you'll get everybody to agree what, or who, is wrong. If you create a moral dilemma and give no identifying information—you strip away any in-group, out-group distinction—you'll get lots of consensus on what's right or wrong. This is what we're finding. But once you plug in the partiality of my group and your group, the entire dynamic changes. And that shows a powerful aspect of the mind.

EM: Well, take the example of the Nazis.

MH: Yes.

EM: This is a question that I still wonder about. They said it's okay to kill Jews: Jews represent a threat to our way of life, to our gene pool, to our values, and so on. They've justified their behavior completely. But, if you think it's okay, then why try to cover it up? Why try to conceal the fact that you're doing it? That becomes the really complex question. You quickly enter this hall of mirrors. You can say, well, they thought it was the right thing to do, but they knew others might not view their behavior that way and that they should therefore cover their tracks. But isn't that tantamount to saying that they knew it was wrong? It's a real question.

MH: It is a real question. But, ultimately, I think it comes back to having a sense of your place in the world. You have a sense of what others will respond to in terms of your actions and, ultimately, that feeds back into your behavior. So I think you're right, both the covering up and the ability to go forward are two parts of the story.

EM: There's a document that I've been really interested in. When the Russians entered Auschwitz in January 1945, the Germans attempted to destroy their records. But Auschwitz was so large that there was actually a building archive separate from the main Auschwitz archive that housed their construction documents. They forgot to destroy it. It was left completely intact.

There is a document in this archive. And written on it with red pencil are notations chastising the writer of the document for using a word for "gas chamber"—*Vergasungskeller*. Essentially, the added notations tell the writer of the document: Don't use that word. You're not supposed to use that word. Don't use it ever again.

So it tells me that there is an investment in hiding, in covering one's tracks.

MH: You seem surprised by that. What's surprising?

EM: I wouldn't say that I'm exactly surprised by it, but it made me think that morality is the combination of two things: "I'm sorry," and "I'm sorry I got caught." There are two things always operating. There's you, and then there's what the world thinks of you.

MH: Yeah.

EM: If I do X, do I feel comfortable doing it? Do I feel comfortable doing X even though I know people will look at me with extreme disapproval?

MH: I mean, that's the categorical imperative, right? If you want to work through the world of rights and wrongs, imagine, would you feel comfortable doing it yourself? And now imagine a world in which everybody would do what you just did, and would you feel comfortable there? You universalize it.

EM: It's a little different. It's not the categorical imperative. It's saying that when I do anything, I have a picture of other people looking at me and possibly disapproving of my actions. Maybe I just want to be liked by people. I don't want people to think that I'm a bad person, or an evil person.

MH: Right.

EM: I do things that may have nothing to do with who I am but a lot to do with how I want to be perceived by others.

MH: Right. And there's the selfish gene view of this, which is that we evolved minds that always take into account the other because it's self-serving, right? I think about what others are going to respond to in terms of my actions because I want to make sure that I'm maximizing my own self-good. And then there's the group selection view, the idea that we act altruistically because it benefits our group.

There are these interesting experiential games that have been developed by economists, and the typical economics view is we're self maximizing, we're selfish, rational players.

EM: Yeah.

MH: So you play these games, such as the ultimatum, where one individual starts off with a pot of money and can offer some proportion of this pot to an anonymous other. For example, let's say I start out with a pot of $10 and I'm told that if I make an offer to someone who I'll never see again, they'll get what I offer and I get what's left over. But if they reject my offer, no one gets anything.

EM: Okay.

MH: This is a one-shot game. On the rational economics view, I should be giving you the smallest proportion of the pot possible. You should accept anything because something's better than nothing, right? But it turns out that most people offer about half, so about $5. And for people who offer less, usually in the range of $1 to $3, the recipient rejects. So there's some notion of fairness, which overrides the selfishness we expect. What would we think of ourselves if we offered so little? How would the recipient perceive that kind of offer? So this all plays into the view that the mind evolved with these regulatory mechanisms that counterbalance complete self-interest.

EM: Yes. I've never been terribly interested in zero-sum games, non-zero-sum games, the game theoretic analysis of human behavior.

MH: What's the intuition, though?

EM: The intuition? To make a crude generalization: In game theory you make a series of calculations. I need to appear generous or fair to ensure that the people around me will be more generous or fair. And so on and so forth. It assumes that people effectively *communicate* with each other.

MH: Say more about that.

EM: Well, very often it seems to me that in trying to analyze human behavior, people create these simplistic games or models that they feel will teach them something. But in real life, you're plunged into the middle of a confusing, uncertain nightmare.

MH: Yeah.

EM: I'll give you an example from Abu Ghraib. One of the soldiers, Sabrina Harman, took pictures of a corpse of al-Jamadi, who was, for all intents and purposes, killed by the CIA. She took pictures of the corpse and pictures were taken of her next to the corpse.

MH: Yeah.

EM: And in one of the pictures, she has her thumb up. People can't get beyond the thumb. Why? They think it tells us she is responsible for al-Jamadi's death. But she's not. The thumb hides the murder.

People look at the picture and make assumptions about what's going on in her head. And they're wrong. They've got it all wrong. And then they express various kinds of disapproval.

MH: Right.

EM: Anger, annoyance, horror. I don't think all of this can be easily analyzed in some kind of psychological model. I get caught up in how people interpret behavior to suit their own interests. How there's this very strange discontinuity between what actually goes on in the world and how the world is perceived; how we're endlessly misperceived by others and ourselves.

MH: Right.

EM: People look at this and they want to know, "How can I explain these pictures? How can I put them into some model of human behavior? How can I square them with my understanding about how human beings act?"

MH: Yeah.

EM: And then they truck out various psychological theories. Theories that I suppose are based on experiment. And it seems to me that the experiments are both misunderstood and don't have the great generality attributed to them. The most obvious example is Philip Zimbardo. Another candidate for this kind of thing is Stanley Milgram. The experiments tell us: People do this bad stuff because people always do bad stuff in such circumstances. It has a kind of circularity.

I'm not sure what these experiments actually show. I'm not sure that they have any application to Abu Ghraib or any other bad situation I can imagine. They often seem to me to be an excuse for not thinking about stuff, for creating a set of blinders around the complexity of the human experience so that one doesn't have to look at it, rather than the other way around.

MH: Right. So I think you've now hit on a true fact about human nature, which is that scientists are always trying to come up with experiments that can account for some significant proportion of variation. Unfortunately along the way we often have a tendency to think we're explaining more of the variation. What I often find happens—and it's happening right here, right now!—is that often when an account of some type of human behavior is brought forth to people not in the sciences, they'll come back and say but here's an example of where that doesn't work. And that's not playing the game quite fairly because the sciences are never going to give a complete explanation of every aspect of human behavior. But the hope is to find some common universal principle that accounts for a significant aspect of our behavior.

EM: Sure.

MH: Now take the Milgram experiments. About a year ago, there was a study done that replicated Milgram's experiment. So you may think how is that possible? Aren't those now deemed unethical?

Well, they are, but we can do them if they're in virtual reality space. This group in London—led by Mel Slater—created the Milgram experiments in virtual reality. So you're the subject and while you're in the experiment, you're hooked up to skin conductance gizmos, which look at the sweatiness of your palms and heart rate and track how revved up you're getting.

EM: Right.

MH: And what you find is that all of the factors that Milgram uncovered in his original experiment—how close you are to the individual, how much you've interacted with him before, how dominant the experimenter is in pushing you forward—all of those get mapped onto the physiological response of the subject in exactly the same way as they did in the original experiment. And they know it's not real. It's like, why do men look at *Playboy* or *Penthouse*? It's just a magazine. But the mind goes on automatic pilot in some cases, blind to reality.

So the interesting thing is that, of course, people know they're in a completely fake environment—it's virtual reality. And yet there are parts of the brain that don't get it. To use a term from cognitive science, there's a sense of encapsulation or insularity, so even though I know this is a visual illusion, I don't give a hoot.

EM: I don't care.

MH: Right. And that says something very important about the moral domain, because there are parts of the brain that are just going to see the world in a particular way independently of rich belief systems.

EM: Right, but wait a second. Tell me what was immoral about the original Milgram experiments.

MH: Oh, well, one thing that has changed since those early experiments is the kind of information that you owe to your subjects when you test them in an experimental setting. That is, there is now a much greater burden to inform subjects of what is likely to happen in the experiment, and some of the negative consequences that might ensue. In the Milgram experiments, the issue was that some people might experience considerable trauma from engaging with the shock device. Thus, there is a burden of informing subjects, right?

EM: What specifically was immoral about the experiments for the subjects, though?

MH: Well, like in many psychological experiments, there was dishonesty. They didn't know that they were not shocking the individual. They were also, in some sense, coerced because the relationship between the experimenter and the subject was asymmetric. And those are the situations that human subject committees are very sensitive to now. You don't want to put people in a situation that forces them to do things that they knowingly don't want to do. You want people to have knowledge of both what they're going to do and the possible consequences of it.

EM: I look at it differently.

MH: Yeah.

EM: For whatever reason. I always thought that the big no-no with the Milgram experiments was that the people who "failed" the experiments, the person who came in and gave the "supposed" heart patient horrifying electric shocks, was going to have to live with that for the rest of his life. You go in there, you take the Milgram test, as it were, and you fail. You flunk.

MH: Right, right.

EM: Instead of saying, "No I won't do it, that's wrong, that's unacceptable," you administer dreadful shocks to people. Then someone tells you what's going on. You may feel lied to, of course, betrayed by the experimenter. But you're still left wondering: Am I a Nazi? Am I a villain? Am I a killer?

I read this book about great experiments in psychology. It was written by Lauren Slater—and very controversial actually.

MH: Oh yes. Right.

EM: And one of them, of course, was the Milgram experiment. They had sealed the records so you couldn't find out who passed, who failed; who eagerly administered electric shocks, who said, "No, I don't want to." She was able to locate the people who had failed the Milgram experiment and interview them. I thought it was really, really interesting. Did you read any of this stuff?

MH: I didn't read it, no.

EM: One person who had failed said, "This was a really valuable experience. Now I see what I'm capable of."

MH: Right. Yes.

EM: I'll be on the lookout. I'll observe my own behavior more closely, more carefully. He talked about it as though he'd received an inoculation. He'd received the Milgram vaccine against genocide. And so I wondered, could there be the Milgram vaccine?

MH: Well, at some level that's what Anthony Burgess plays with in *A Clockwork Orange*, right?

EM: Maybe.

MH: Some of those behavioral therapies have had success, to varying degrees. You know there was a program on—my great source for knowledge of the world—MTV.

EM: It's probably a really good source.

MH: Ha—right. It was a good source. It was a program on these intervention studies being done with kids who are juvenile delinquents who have been convicted of some fairly horrible crime and are then put into a prison where some of the more horrible convicts show them about the brutality of what it's like to be in a prison. Because a lot of these kids don't have any idea. And apparently these interventions are relatively successful, and with many of these kids they actually have a defeating effect.

EM: Interesting you should mention this because there's a film called *Scared Straight!* It's one of the worst things I've ever seen. It may have gotten an Oscar for Best Documentary Feature.

It was the same idea. Kids who had committed various crimes were jailed with hardened criminals in a maximum-security prison. And they're "scared straight."

MH: Right.

EM: Well, the movie was really, really bad. On the level of craft, I thought:, "This is just awful." Then someone wrote an article about it. The kids in this program had come from an upper middle class suburb where the recidivism rate is zero. So, now I thought: "Bad movie *and* really bad social science." If the recidivism rate is zero, then it really doesn't matter whether they are "scared straight" or not, they are going to "go straight" regardless. And then I thought: "Really bad morality." Is this how you want to ensure compliance in a society? By scaring people?

The Milgram inoculation is a little bit different. It's scaring people by giving them an example of what they're capable of. It's not by saying, I'm going to take a two-by-four to the side of your head. It's not *A Clockwork Orange*.

MH: See what was more interesting about the MTV version of the documentary was that they actually talked about the cases that did not work. And those cases—where going to the prisons and scaring the pulp out of the kids had no effect—were cases where macho, violent behavior was part of their culture. And if the kid went back and said, you know, "I'm cured," the kid had less of a chance of surviving. And so, in the same way that a Milgram-type inoculation will work for some people, a scared straight version will work for other people.

EM: Okay.

MH: There was a study done a couple of years ago with a very, very large sample of boys; something like five hundred boys who they'd tracked from a young age into their juvenile years. What they found was a gene with two modes of expression, so to speak. One mode expressed a high concentration of neurochemicals, like serotonin and dopamine, chemistry linked to our impulsive behavior. The other mode expressed a low level of these chemicals.

EM: Uh-huh.

MH: Now, here is the fascinating part. Among the sample of boys observed, there were some who led a wonderful childhood, had caring parents. A second category was subjected to mild levels of violence from their parents. And the third category received high and fairly abnormal levels of parental violence. When you look at the profiles of these children during their juvenile-teen years, you find a striking pattern. The boys with the high expressing form of the gene appeared to be buffered against parental violence, whereas the boys with the low expressing form weren't, and they showed far higher rates of juvenile delinquency.

EM: Really?

MH: So it's a beautiful, albeit sad, case of gene-environment interaction. Depending on the cards you get, and the environment you land in, you could either end up fine, because you are genetically buffered from the nasty environment thrown your way, or deeply injured by this environment because there are no genetic defenses. This shows the importance of looking at how genetic systems build bodies and minds that set up opportunities for the environment, opportunities to sculpt outcomes in particular directions.

EM: Okay.

MH: I mean, it's this kind of genetic view that indicates why we're having this conversation, but dogs aren't, you know? After all, dogs are exposed to pretty much the same environment we are, but we chat, gossip, and pontificate, whereas they bark, with a few variants here and there.

EM: Well, as far as we know dogs aren't having this conversation.

MH: As far as we know, right. As Gary Larson put it, when we listen in on dog conversations, all we hear is "heh, heh, heh."

EM: Right, exactly.

MH: But this study is a nice example of how developmental factors can greatly influence morally relevant behavior in later life. This kind of story may have important implications for psychopathy where there is increasing evidence for early developmental problems, well before the signature of psychopathy shows itself in early adulthood. It's now fairly well documented that people who become psychopaths show signs of this disorder early, in the form of violence toward pets. Then, later in life, this violence migrates to humans.

EM: From their pets? I'm suspicious of the whole idea of psychopathy.

MH: You can't go there.

EM: No, no, no. We *can* go there. Why not? ∞

Seventeen

Free Will
Tom Wolfe and Michael Gazzaniga

*The original New Journalist and the father of cognitive neuroscience
discuss status, free will, the human condition, and the Interpreter.*

Tom Wolfe, who calls himself "the social secretary of neuroscience," often turns to current research to inform his stories and cultural commentary. His 1996 essay, "Sorry, But Your Soul Just Died," raised questions about personal responsibility in the age of genetic predeterminism. Similar concerns led Michael Gazzaniga to found the Law and Neuroscience Project. When Gazzaniga, who had recently published *Human: The Science Behind What Makes Us Unique*, came to New York City in 2008, *Seed* incited a discussion on status, free will, and the human condition.

Novelist and journalist Tom Wolfe spent his early days as a *Washington Post* beat reporter, where his free-association, onomatopoetic style would later become the trademark of New Journalism. Wolfe's 1979 book *The Right Stuff*, an account of the pilots who became America's first astronauts, won the National Book Award for nonfiction. In that work, and in others such as *The Electric Kool-Aid Acid Test* and *The Bonfire of the Vanities*, Wolfe delves into the eccentricities of human behavior, language, and social status with a mastery many consider unparalleled in the American literary canon.

Considered the father of cognitive neuroscience, Michael Gazzaniga has long been fascinated by how the mind emerges from the brain. His work, which focuses on patients who have undergone split-brain surgery, underpins his many books, including *Nature's Mind, The Ethical Brain*, and his latest, *Human: The Science Behind What Makes Us Unique*. At the University of California, Santa Barbara, Gazzaniga serves as director of the cross-disciplinary SAGE Center for the Study of the Mind. He also served on the President's Council on Bioethics.

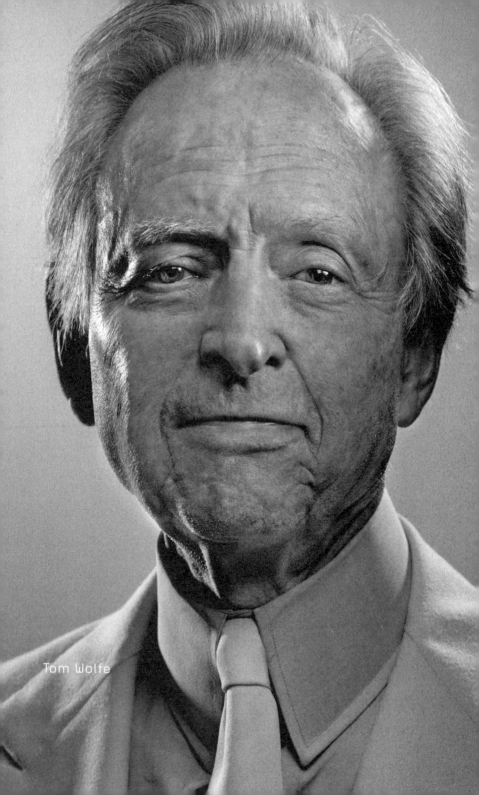
Tom Wolfe

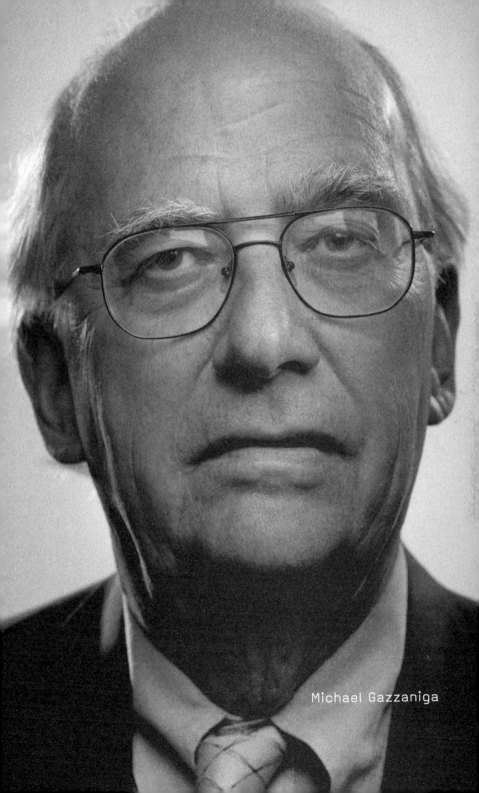
Michael Gazzaniga

Tom Wolfe: Mike, I don't want you to think I'm giving up my right to disagree with you down the line—I may not have to—but you're one of the very few evolutionary thinkers and neuroscientists that I pay attention to, and I'll tell you why. In the nineties, when the subject of neuroscience and also genetics started becoming hot, there was a tendency to conflate genetic theory and evolutionary theory with neuroscience, as if the two were locked, which just isn't true. Remember José Delgado, the wave brain physiologist who was at Yale at one time?

Michael Gazzaniga: Oh yeah. Sure.

TW: The guy stood in a smock in a bullring and put stereotaxic needles in the brain of a bull and just let himself be charged. He had a radio transmitter. The bull is as far away as that wall is from me, and he presses the thing and the bull goes *dadadada* and comes to a stop.

MG: Right.

TW: He's still with us; he's in his nineties. Anyway, his son, also José Delgado, and also a neuroscientist, was interviewed recently and he said, "The human brain is complex beyond anybody's imagining, let alone comprehension." He said, "We are not a few miles down a long road; we are a few inches down the long road." Then he said, "All the rest is literature."

Many of today's leading theorists, such as E. O. Wilson, Richard Dawkins, and Dan Dennett, probably know about as much on the human brain as a second-year graduate student in neuropsychology. That isn't their field. Wilson is a great zoologist and a brilliant writer. Dawkins, I'm afraid, is now just a PR man for evolution. He's kind of like John the Baptist—he goes around announcing the imminent arrival. Dennett, of course, is a philosopher and doesn't pretend to know anything about the brain. I think it has distorted the whole discussion.

MG: Well, let me roll the cameras back to the eighties and nineties, when neuroscience was taking off. There were new techniques available to understand the chemical, physiological, and anatomical processes in the brain. Imaging was starting up and the inrush of data was enormous and exciting. So there was a hunger for the big picture: What does it mean? How do we put it together into a story? Ultimately, everything's got to have a narrative in science, as in life. And there was a need for people who didn't spend their time looking down a microscope to tell a story of what this could mean. I would say that some of the people who've made attempts at that did a very good job. But I will hold out for the fact that if you haven't slaved away looking at the nervous system with the tools of neuroscience—if you're only talking about it—you don't quite have the same respect for it. Because it is an extraordinarily complex machine. If José Delgado says we're two inches down the road to this long journey, I would say it's more like two microns.

TW: Right.

MG: It's a very daunting task. When I was at Dartmouth College in the late fifties studying biology, they were just beginning to tell us about DNA. It was a dream. Linus Pauling said, "Someday there's going to be molecular medicine." And the response was "What are you talking about?"

In the past fifty-five years, there's been this explosion of work and incredible, intricate knowledge about how genes work. My youngest daughter is now a graduate student in genetics, I'm happy to report. So this past Christmas, I said, "I'm going to buy a genetics textbook and read the sucker, and I'm going to be able to converse with my daughter." I got to page two, and I said, "I'm going to talk to her about other things."

TW: Ha ha.

MG: It's far too complicated. But it's at a point where there's an explosion of information all over the world. And you feel it—the next new idea is waiting to happen.

TW: I think all this excitement has spawned a replacement for Freudian psychologists. They've been replaced by the evolutionary psychologists, whose main interest seems to be to retrofit the theory of evolution on whatever ended up happening. I read an example in your new book of a woman who's come up with an elaborate theory that music has a survival benefit in the evolutionary sense because it increases the social cohesiveness of populations. I would love for her to read a piece that appeared recently in *The New Yorker* about a tribe, the Pirahã on the Maici River, a little tributary of the Amazon. This tribe, it turns out, has a language with eight consonants and three vowels. I think they have a sum total of fifty-two words or something like that. As a result, they have little art, they have no music, no dance, and no religion. They're usually cited because they seem to be a terrible exception to Noam Chomsky's rule that all people are born with a structure that enables them to put words in a grammatical form. Not the Pirahã! And they're not stupid or retarded in any sense. They just had never increased their language abilities—and they don't want to.

MG: Yeah. Well, exceptions are historic. Look, the good evolutionary psychologists are good. They're telling us not to fall into the trap of thinking that everything's fixable via simple learning mechanisms or social engineering. They're saying, "Look, there are basic aspects to human nature that are common to all members of our species and have been there a long time." What's exciting is that we've developed this cognitive mechanism to free us from the things that determine so much of our behavior. And by doing so, we've sort of cut the rope from the rest of the animal kingdom. We can do things and we can cultivate certain behavior, even though there are obviously a lot of tendencies that are part of our biology. For example, here's an idea that comes from evolutionary psychology, an observation that I think is rather shrewd: Why are members of our species drawn to the fictional experience? Here you are, someone who's spent your life with fiction—

TW: I was at one time a journalist. We don't deal with fiction. Not intentionally.

MG: Ha ha—right. But it's a fascinating thing to think of the role that fiction and make-believe play. Do you feel, when you create a body of fiction, that you're opening up possibilities for people to think about problems in a different way? To confront things they don't yet know about?

TW: Well, I do take issue with the idea that all stories have a bearing on evolutionary benefits or survival benefits. In my opinion all stories have to do with status. When people say, "I just want some good escape literature," what they're looking for are dramatizations of people facing status problems. Harry Potter is like every child who feels overwhelmed by this adult world around him, and he overcomes it in ways that don't interest me in particular—he can pull things out of the air. But, like *Anna Karenina*, it's a story about status problems. Tolstoy and Flaubert would be paupers today, writing these novels, which are all based on the idea that a woman must remain chaste. They'd be laughed out of town. The story of Anna Karenina and Vronsky would be a Page Six item and then that would be the end of it. But if we successfully put ourselves in the mind-set of the nineteenth century, we can really enjoy the status problems that they have.

MG: Do you think all art is about status?

TW: Well, certainly not music. Dance, maybe yes, maybe no. But literature and movies, yes. To me the crucial point is something, which I don't think even Chomsky understands, about speech and language. Chomsky and many other people are wonderful at telling us how language works, and about differences in languages and the historical progression of languages across the face of the Earth. But I seem to be the one person who realizes the properties of speech. Speech is an artifact. It's not a natural progression of intelligence, in my opinion—we have to look only at the Pirahā for that. It's a code. You're inventing a code for all the objects in the world and then establishing relationships between those objects. And speech has fundamentally transformed human beings.

MG: By speech I assume you mean language and not the actual act of speaking?

TW: To me, it's the same, speech and talking.

MG: Okay, so what do you think language and speech are for? I mean, it's probably an adaptation. We're big animals, and that's one of the goodies that we got.

TW: I think speech is entirely different from other survival benefits. Only with speech can you ask the question "Why?"

MG: Right.

TW: Animals cannot ask why. In one way or another, they can ask what, where, and when. But they cannot ask why. I've never seen an animal shrug. When you shrug, you're trying to say, "I don't know why." And they also can't ask how.

MG: Yeah.

TW: With language you can ask that question. I think it's at that point where religion starts.

MG: Right.

TW: Humans got language and they were suddenly able to say, "Hey, why is all this here? Who put it here?" And my assumption is that they said, "There must be somebody like us but much bigger, much more powerful, that could make all these trees, the streams. God must be really something, and you'd better not get on the wrong side of him." I think that's the way it started.

MG: As you may know, I came across this phenomenon that I call the Interpreter. It's something that's in the left hemisphere of the human; it tries to put a story together as to why something occurred. So we found this in patients who've had their brains divided. What we could do is sort of tiptoe into their nonspeaking right hemisphere and get them to do something like walk out of the room or lift their hand up. Then we would ask the left hemisphere, "Why did you do that?" And they would cook up a story to make sense out of what their disconnected right hemisphere just did. The left brain didn't know that we'd pulled a trick on them, so they concocted an explanation for why they walked out of the room. And it's because this left hemisphere can ask, "Why? What's that all about?" But one of the things we've never been able to unpack is whether this Interpreter is completely overlapping with the language system and is therefore a sort of press agent for its own mechanism. What we do know is that there are separate systems for different types of cognition. And the Interpreter seems to be located in the parts of the brain where language is located. So many people do think that interpretive capacity comes with language; that this is the deal with language—it comes along for the ride. Others believe that there are actually all kinds of different cognitive mechanisms happening, and language reports them out. So the function of language is to talk about it, talk about what you know and communicate, "Hey! Look here, I know how to cook a fish. Here, let me show you how."

TW: I've always been interested in your theory of the Interpreter. When I was in graduate school, I was introduced to this concept of social status in the work of Max Weber, the German sociologist. And the more I thought about it, the more I could see that status was not simply something that was appearances and houses and automobiles, or even ranks in a corporation or that sort of thing. It invaded every single part of life. I remember when I was in graduate school, there was a setup wherein a common bathroom was shared by two rooms. And there was a student from India—a brilliant scientist— who had apparently come from way out in the countryside, with no natural social standing and not many amenities. Now, you'd think the things you do in absolute private would not be driven by status concerns. But he heard three of his American friends joking about the fact that when they went into the bathroom, they found footprints on the toilet seat. Well, this fellow had never seen a porcelain toilet before. He was crushed. He felt absolutely humiliated, and here was something that goes on in private.

Anyway, this was before I'd ever heard of neuroscience, and I said, "There must be something in the brain that registers this, your status in every kind of situation." And I kept looking for it. Freud had been such a powerful figure that everyone seemed to think, "Freud's got the bottom line, why should we go through all these complicated neurons and everything to see how he got there. He's got it." I hoped to find the answer in Delgado's book, but it wasn't really there. It wasn't until I ran across your concept of the Interpreter that I thought, "Hey, maybe we've got it."

MG: Well, the key concept in understanding status has to do with the idea of social comparison. The Interpreter fires up and almost reflexively starts to compare the new person with one's self and others. Multiple factors seep into this narrative being built by the Interpreter and the importance of status is one of the products of that process.

Still, I think the essential question that neuroscience has to answer is why, when I interact with someone, I don't think it's my brain talking to their brain. I'm talking to Tom Wolfe, and you're listening to Mike Gazzaniga, right? We instantly convert to that: I give you an essence right off the bat. I put you at the level of a person with mental states and all the rest of it. That mechanism, it makes us all dualists in a way. Absolute dualists. That mechanism is the deep mystery of neuroscience, and no one has touched it yet. No one knows how that works. That's the goal.

For my part, what I've come to realize is that the neuroscience of the next twenty years will be studying social processes of humans. In order to get to the biology of anything, you need technology that allows you to study the human mind. It's only really in the past ten or fifteen years that we've had the new methods of imaging. And they keep getting better and better and better. The ability to think about other people is probably the impetus behind all these marvelous things the human brain can do.

TW: Every time we go into a room with other people, it's as if we have a teleprompter in front of us and it's telling us the history of ourselves versus these people. We can't even think of thinking without this huge library of good information and bad information.

MG: That's why the great psycholinguist George Miller, whom we shared a dinner with once, called us the "informavores." That's how he wanted to cast us.

When you get up in the morning, you do not think about triangles and squares and these similes that psychologists have been using for the past one hundred years.

You think about status. You think about where you are in relation to your peers. You're thinking about your spouse, about your kids, about your boss. Ninety-nine percent of your time is spent thinking about other people's thoughts about you, their intentions, and all this kind of stuff. So sorting all that out, how we navigate this complex social world, there's going to be a neuroscience to it, and I think it's going to be very powerful.

MG: I'm involved in a new project called Neuroscience and the Law, which I think you're familiar with. It brings up the idea that there are these causal forces that make us do the things we do, that by the time you're consciously aware of something, your brain's already done it. How else could it be? Because the brain is what's producing these mental events that we're sorting through. So these ideas— what I call the ooze of neuroscience—are going out everywhere, and people are willing to accept that: "My brain did it. Officer, it wasn't me." These defenses are popping up all over the judicial system. But if we adopt that, then it's hard to see why we have a retributive response to a wrongdoing. It would seem to me to be morally wrong to blame someone for something that was going to happen anyway because of forces beyond their control. So people get into this loop, and they get very concerned about the nature of our retributive response. This puts you right smack in the middle of the question: Are we free to do what we think we're doing?

TW: Oh, it's the hottest subject in academia. Philosophy students are flocking to neuroscience because they think the answers are all there, not in our silly, cherished way of thinking. It's called materialism to some. We are computers, and a computer is programmed a certain way, and there's nothing the computer can do to change its programming. I think materialism is too grand a word for it. It's mechanical. I mean, here's what happens. The scientist says, "We are machines." There's no ghost in the machine. There's no little tiny "me" in the conning tower surveying the universe and figuring out a place within it. It's a machine. Things get more and more complicated when it comes to humans, but it's still a machine. Obviously, this machine has no free choice. It's programmed to do certain things. It's as if you threw a rock in the air, and in midflight you gave that rock consciousness. That rock would come up with twelve airtight, logical reasons why it's going in that direction. This has caught on like wildfire. The flaw in that is that speech, language, creates so many variables. Speech reacts. It's the only artifact I can think of that reacts.

MG: Well, I think using the term "free will" is just a bad way to capture the problem. Because here's the question: Free from what? What are you trying to be free from?

TW: It's a very simple definition: You make your own decisions.

MG: Yeah. But who is "you"? "You" is this person with this brain that has been interacting with this environment since you were born, learning about the good and the bad, the things that work and don't work. You've been making decisions all the way along, and now you have a new one and you want to be free to make it. So psychologically, the Interpreter is telling you you're making this decision. But the trick is understanding that your brain is basing the decision on past experience, on all the stuff it has learned. You want a reliable machine to make the actual act occur. You want to be responding rationally to any challenge that you get in the world, because you want that experience to be evaluated. That's all going on in your brain second by second, moment to moment. And as a result, you make a decision about it. And phenomenologically, when the decision finally comes out, you say, "Oh, that's me!"

TW: Speech has introduced so many variables into your machine that it becomes pointless to argue whether this is free or not free will. Obviously, it's not free in the sense that if you don't have this body, you can't do anything. But it is free in the sense that because of your experiences and because of the reactions of speech constantly feeding you new material, your brain is going to operate differently from anybody else's, and that is the free will—whether you call it mechanical or not. Everybody becomes such an individual, it becomes pointless to say, "You didn't make that decision." It's an absurd idea.

MG: Well, I think we're saying the same thing. There is a very clever little experiment that you would be amused by, run by my colleague Jonathan Schooler. He has a bunch of students read a paragraph or two from the Francis Crick book *The Astonishing Hypothesis*, which is very deterministic in tone and intent. And then he has another group of students reading an inspirational book about how we make our own decisions and determine our own path. He then lets each group play a videogame in which you're free to cheat. So guess who cheats? The people who have just read that it's all determined cheat their pants off.

I think people who try to find personal responsibility in the brain are wrongheaded. Think of it this way: If you're the only person in the world, you live alone on an island, there's no concept of personal responsibility. Who are you being personally responsible to? If somebody shows up on the island though—

TW: Friday was his name.

MG: Yeah, exactly. Then you've got a social group. And the group starts to make rules; that's the only way they're going to function. Out of those rules comes responsibility. So responsibilities are to the relationships within the social groups, and when someone breaks a rule, they're breaking a social rule. So don't look for where in their brain something went wrong; look at the fact that they broke a rule, which they could have followed. I'm actually kind of hard-nosed about this. I think people should be held accountable for lots of stuff.

TW: No, I would certainly agree with that. In fact, my theory of status is that all of us live by a set of values that, if written in stone, would make not me but my group superior in some way. I think there are just so many kinds of status layers due solely to likeness. You can always find a group that seems to justify whatever you're doing.

MG: Our species seems brilliant at forming groups—indeed support groups—for almost anything. And no matter what the group is about, no matter what its character, it becomes advocatory. ∞

Tom Wolfe and Michael Gazzaniga

Eighteen

Evolution, Creativity, and Future Life
Jill Tarter and Will Wright

The astrobiologist and the game developer discuss model-making, the singularity, and the value of scientific revolutions.

Will Wright's lifelong interest in astrobiology deepened during his numerous visits to the SETI Institute over the evolution of *Spore*, the video game Wright designed. There he became familiar with the work of Jill Tarter and her colleagues, whose mission to explore the universe for signs of life inspired Wright's development of the game. Their conversation in Manhattan was characteristically ambitious, raising such questions as, Can we model reality? How do we quantify scientific revolutions? And is the singularity inevitable?

A pioneer in the field of astrobiology, astronomer Jill Tarter has dedicated her career to the search for extraterrestrial intelligence. As director of the Center for SETI Research at the SETI Institute, Tarter seeks radio signals from alien civilizations using the Allen Telescope Array. Her work inspired Jodie Foster's performance in the movie *Contact* and has garnered many accolades, including two public-service medals from NASA. In 2004, *Time* named Tarter one of the most influential people in the world, and in 2009 she received the TED Prize and the opportunity to make a wish to change the world.

Will Wright designs computer games. In 1987 he co-founded the videogame development company Maxis, which created *SimCity*, a game inspired by theories of artificial life, human psychology, and urban planning. *SimCity* ultimately spawned an entire genre of "Sim" games, notably *The Sims*, which is the bestselling PC game in history. In 2008, Wright released *Spore*, an ambitiously scaled, science-inspired "life simulation" game. Inventor, tinkerer, and entrepreneur, Wright was inducted into the Academy of Interactive Arts and Sciences' Hall of Fame in 2002, and in 2005 was awarded *PC Magazine*'s Lifetime Achievement Award.

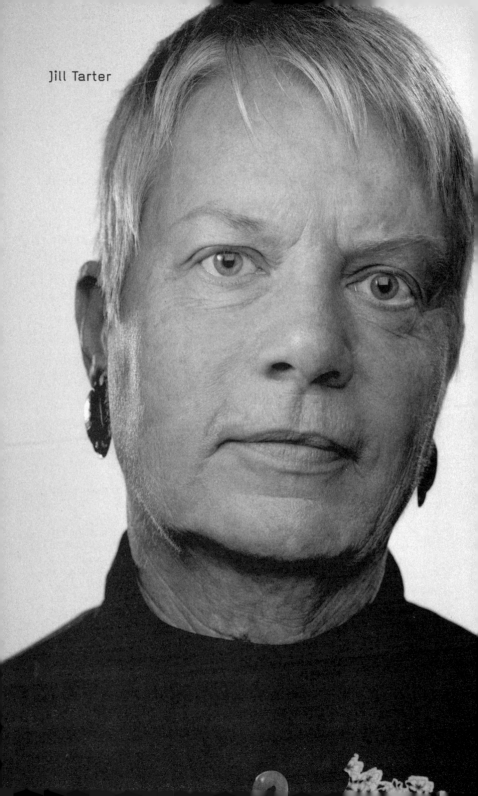

Jill Tarter

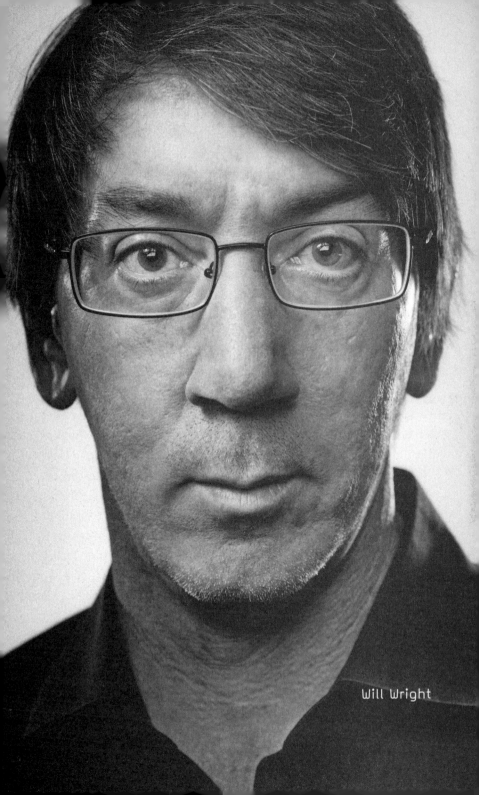

Will Wright

Jill Tarter: I'm wondering if *Spore* is setting the stage for the next generation. Should we become more machinelike? Should we develop ways to think and evolve that biology didn't give us?

Will Wright: Well, I think games and the technology they use are tools we can layer on top of what we've already got. The human imagination is this amazing thing. We're able to build models of the world around us, test out hypothetical scenarios and, in some sense, simulate the world. I think this ability is probably one of the most important characteristics of humanity.

JT: One of the things about games is that you get to build a model without any consequences or constraints. So *Spore* in one sense is a very good way of looking at evolution. But, if all things are possible with a certain number of DNA points, and you're not constrained by the physical world or what that world would actually provide you, then where do you learn about the consequences of evolution?

WW: Well, when kids—or adults—play a game, there's a model in the computer that they're playing against. And when they play they're reverse-engineering that model. As they get better at the game, they get a more accurate representation of that computer model. And what I've seen in almost every game I've made is that when players get a close representation of the model they're playing against, they transcend it. They start arguing with the assumptions of that model, saying, "Hey, I don't think that's the way cities really work. I don't think mass transit's really that effective." And when they surpass the computer model, it feels to me like they're graduating the play experience—they realize this is a toy, not an accurate model of reality.

JT: Okay, so now they've got a better idea. How do they put it into effect, if it's not already built into the structure of the game?

WW: Well, at that point it becomes really interesting, because if a lot of people are playing the same simulator, they can now have an intelligent debate about how they think it differs from the way the world really works. I think, in some sense, being able to use this imaginary model as a shared landmark helps them actually increase the resolution of their model of the world around them. So building that model in the first place, I think, is a great path to put people on.

JT: I agree with you. But, again, I'm eager to understand how learning to be good at a game makes you good at life, makes you good at changing the world, and gives you skills that are going to allow you to reinvent your environment. Because, in the game, you play against an environment that's been given to you.

WW: I don't think of games as something to replace traditional education. There's the saying that education is not the filling of a pail, but the sparking of a fire. If you can spark an interest in a kid, then you just have to get out of the way. Very frequently really cool science is hidden beneath layers of academic language and terminology. And I think things like games and entertainment in general show kids why these subjects are fascinating in a language that they can understand.

So with *Spore* we took a lot of artistic license in doing that. For example, when you get out to the space level, you have intelligent civilizations all around you.

JT: Right.

WW: Which, as I understand, we're not observing, though hopefully we will some day. But if I put you in an empty galaxy with nothing but microbes that would have been a far less engaging game. As it is now you have a starship, and usually within twenty or thirty stars' distance out, you'll find some other species.

JT: That's a lot.

WW: Yeah, I thought you would agree that's an obvious exaggeration.

JT: Well, mainly because the twenty or thirty stars closest to us are M dwarfs and not likely to be good candidates for hosting. Also it doesn't show how empty space is, how far it is between things. Even with a lot of intelligent life, we could be functionally alone because there's no way for us to communicate over the time scales that we have available to us.

ww: Right. So, again, we're taking a huge amount of artistic license to keep the game engaging and interesting. But I think motivating people to understand, to come back and say, "Wait a second, that's not right" or "That's not reasonable given the evidence" is a good thing—when players realize their own model of reality is superior to this toy model that we've given them.

JT: I keep thinking about the generation that's getting exposed to all this wonderful, rich opportunity of game-playing as education, and that they expect to be able to manipulate the real world the way they do the game world. How do we bridge that? How do we turn them into socially functioning members of humanity on one planet?

ww: It's funny, because I think they are able, more and more, to manipulate the real world like the game world. If you look at the tools that they have available on their cell phones, Google Maps, and such, the amount of formalized information that we can extract from the world around us is skyrocketing. And it's very much based upon things like game interfaces.

JT: But the fact that they can use that interface and pull up this information hasn't changed one iota of the information content. They're just accessing it.

ww: Oh, they're consumers of it, correct. Although more and more, they do have the ability to produce it, to broadcast a video on You-Tube or their Web pages, et cetera.

JT: Right. But this takes me back to what we're doing as we use games to study evolution. I mean, are you, Will, the great Pied Piper who is leading our kids into a future where they will accept enhanced attributes in, or on, their own bodies and give up some of the biological aspects of humans as we know them now? Are you leading the way to the singularity?

ww: Ooh, that's quite an accusation!

JT: I'm asking a question. I'm not even saying it's a good thing or a bad thing.

WW: Transhumanism, you mean?

JT: Yeah. Are these kids being primed to accept that? To want that? To try to make it happen?

WW: Well, as I said, if there's one aspect of humanity that I want to augment, it's the imagination, which is probably our most powerful cognitive tool. I think of games as being an amplifier for the imagination of the players, in the same way that a car amplifies our legs or a house amplifies our skin. Not only are we able to build much more elaborate models on a computer, which can keep track of all the numbers and the repercussions, but we're also able to share and communicate those models to others. It becomes a tool of self-expression.

JT: Right.

WW: On a somewhat deeper level is the idea of proving to people that they can be creative. Because when you're in this world, seeing all these incredible movies and books and whatever thrown at you, the idea that you as an individual can be creative is almost educated out of you.

JT: Yeah, you don't have to teach a three-year-old to be creative; they're as creative as hell.

WW: What I want to do is take the natural motivation people have to play games and put it toward some amount of reality. Even if it's a toy reality at least they can see, "Oh, this is about evolution and astronomy; it's about culture." So that at least at a symbolic, simple, toylike level, we're representing the world around them, as opposed to an obvious fantasy world of Orcs, and—

JT: Magic wands.

WW: Right. Basically replace the magic with science, because I think science is every bit as magic as any magic. I never really like fantasy books for that reason. I was never a big Tolkien fan. I always felt that magic kind of removed the drama from a story because magic could do anything; there were no limitations. It's like what you were saying earlier, about fundamental limitations that bring you back to reality, that ground you.

In most games there's a more or less obvious goal you can shoot for and there's some way to achieve that goal. In science, we don't know that. For the science you're pursuing, it's not apparent that the stated goal is achievable, right?

JT: Correct. That's why it's a question.

WW: Yeah. And it's a very interesting one. I think how you value that science is very difficult for a lot of other people, since you might never succeed. And even if you succeed, what does that mean? I'm sure it's going to change the world in huge, unimaginable ways if you hear from somebody. But typically, that kind of science is presented as a return-on-investment thing, right? If we achieve X, it's worth this much, so if it's going to cost us that much with this probability . . . You run the numbers and say, let's invest in it or not.

JT: There are only a very few people and only a small amount of money invested in my project, so yes, it is about ROI, but not necessarily the traditional kind.

WW: Have you ever tried to quantify the return? Assuming that you did hear from somebody—and it seems like there may be two or three major possibilities. Say we hear from someone that we know is intelligent, but we can't decipher the message, we don't understand them.

JT: Cosmic dial tone.

WW: Yeah.

JT: So that's a proof of existence.

WW: Have you tried to actually quantify what that would mean?

JT: No, I can't quantify what the knowledge that it's possible to have a long future as a technological society is really worth. Because presumably I'm not smart enough to do that. But that's the underlying message that a success provides. Not information transfer, not extraterrestrial salvation. It's the fact that success is impossible without technology; two short-lived technologies are unlikely to find each other during the ten-billion-year history of the galaxy. Successful detection means that the longevity of technological civilizations is the rule rather than the exception.

WW: Right.

JT: So right now we don't necessarily see a long future for technological civilizations on this planet. We see so many ways that it could go otherwise. I think a proof of existence, that at least someone somewhere survived their technical infancy, is very valuable, but I can't quantify it.

WW: It seems like an interesting thought experiment, just to figure out how you would attach a value to that. You'd have to start with cultural awareness, probably in the form of a very long-term reformulation of the way people on Earth think about the future. If you had to resolve that into dollars I'm sure it would be worth trillions and trillions to us. And getting back to the ROI, if you say, "It's going to end up costing us this much and it's a one-in-a-million chance"—if it's times a trillion you can actually put numbers behind it.

JT: So what about the Darwin revolution, or the Copernican revolution, what are they worth in terms of dollars?

WW: Ooh. Well I tend to think of the Copernican revolution as something that helped us think outward with clearer eyes. And of Darwin as almost the opposite. We looked inward to life itself.

JT: Both very valuable, though.

WW: Oh yeah.

JT: But how valuable?

WW: How do you quantify that? To start with Darwinism, I think there was definitely some impact on Watson and Crick, which accelerated our understanding of genetics. Which, in fact, should be quite quantifiable via the benefits we get through modern medicine, biotechnology, et cetera. That one seems more tangible than the

Copernican one. Although the Copernican one, I think, probably accelerated our understanding of astronomy to the point where we had a clearer view of things like our planet, how it works, leading to space flight—

JT: Geosynchronous satellites. The communication net.

WW: Yeah. It seems like you can eventually resolve those to what makes my cell phone work.

JT: Okay. So then what's the value of holding up a mirror and showing the Earth as a planet with one species on it, that species being intrinsically different from the species you just got the dial tone from? How important is it to trivialize difference among human factions?

WW: Yeah, I would hope that would be more of a political impact.

JT: Well, politics is costly, if not valuable, that's for sure.

WW: I mean political in the sense that, if you asked somebody, "What are you?" the most common answer was "I'm human," as opposed to "I'm American," or Peruvian or whatever. I think there'd be a huge value if it meant people first and foremost thought of themselves as earthlings or humans. Because all of the really tough problems we're facing now are planetary problems.

JT: Sure. Cell phone communication: a certain amount of dollars. Bioengineering: another amount of dollars. A future for the planet: priceless.

WW: Totally. There's real value in being pushed toward global awareness and looking long-term. That's one of the things that I find very useful about games. When kids play *SimCity*, they're actually seeing the evolution of a city over decades in a matter of hours. So they see very long-term ramifications of tax policies and land-use decisions playing out before their very eyes. Where in reality, a politician will make these decisions and be out of office, probably even dead, before a lot of these policies really have their major impacts. I think these are the timelines we need to be looking at—the one-hundred- or two-hundred-year horizons. Because most of the really bad stuff that's happening right now is the result of very short-term thinking.

JT: Seems we can hardly do any other kind.

WW: Do you think people as we recognize them will exist in ten thousand years?

JT: Yeah, I think they will be recognizable. I don't think they're going to be like us.

WW: You mean you would actually recognize them as our descendants?

JT: In the sense that we've had a lot of opportunity, through science fiction, to think into the future, and to think about alternate evolution, alternate possibilities. And I think we haven't written anything that we can't imagine, so if we participate, what will happen is probably imaginable.

WW: Because one of the things that has always captured my imagination, especially recently, is intelligent machines and what they're going to mean to humanity. Whether we end up absorbing them within ourselves, whether they become an external species, if you will, going off on their own trajectory. Or whether we both end up diversifying to where there's every combination imaginable—pure mechanicals, pure biologicals, and everything in between.

JT: Why would you keep the pure biologicals around? They think really slowly.

WW: Humans do a lot of strange things. I can imagine a strain of humanity deciding to remain pure, whatever that means, and resisting all subsumption of mechanical augmentation. And maybe they even become Amish or something, living on little reservations.

JT: But they live for only a little while, relative to all these other things they're going to create.

WW: I understand, I'm just saying that human culture is amazingly diverse. And if you look at people's beliefs and span the gamut of possibilities, I would imagine that's one scenario that is probably going to exist in some form. Life in general tends to diversify, given every opportunity. And if we achieve, say, mechanical AI, I see no reason that it would be a homogenous, borglike intelligence. Why wouldn't it diversify every bit as much? Or more so, because we'll have opportunities to create diversity. So we'll have radically different mechanical intelligences with radically different motivations behind them. But at the same time, we'll probably have almost every imaginable combination of humans and mechanicals as well.

JT: It's the slow-thinking speed of the humans relative to the mechanicals that makes me wonder whether in fact they might be so reduced in number that a small glitch, a fluctuation, can mean the end of them. Or whether they won't be purposely eliminated, or choose to eliminate themselves.

WW: It's an interesting point, because if you basically take the brain out, what other part of human biology would you want to keep?

JT: Right.

WW: In which case, is that to say that we're likely to transition into purely designed creatures or entities?

JT: That's what the singularity is all about—having the first opportunity ever to do that. Who will take it is the question.

WW: If that occurs—and it seems likely—then it's certainly going to occur within ten thousand years. Probably within one thousand. Maybe even within two hundred.

JT: And that exponential is just amazingly robust. But I think we would still recognize them. I think we would still see them as having lots of attributes of life.

WW: But I could easily imagine these things being strange in form—an intelligent puddle that thinks in an entirely alien way. And I might just as well think it's an intelligent alien as a descendant of humanity.

JT: Oh, I see what you're trying to get at. Would we claim it as our descendant? I don't know. Do we recognize ourselves as a descendant of some of the really, really weird early species?

WW: No.

JT: We are.

WW: If you look at a genetic basis, yeah. We share so much with a fruit fly it's ridiculous.

JT: But that's the biological that's going to go away potentially.

WW: Right.

JT: So would we recognize it?

WW: I would contend that we are probably more similar to a fruit fly than these descendants will be to us.

JT: And a fruit fly doesn't recognize us, but we do recognize a fruit fly. And we can appreciate the inheritance.

WW: That's interesting. So they'll recognize us as their progenitors, but we won't recognize them as our descendants.

JT: It could be a possibility.

WW: I think you're dealing with two very fundamental questions in your work. Number one, What is the definition of life? And number two, What is the definition of intelligence?

JT: Actually, I don't deal with either. Life, I assume, is a precursor to some technology. And the technology or the intelligence is something that modifies its environment in ways that we can sense, with our emergent technology, over interstellar distances. I really don't care for any more profound definition than that. I'm just very pragmatic.

WW: So when you're looking at signals, you're basically sorting what would be considered an intelligent signal from a natural signal.

JT: Right, an engineered signal from a natural signal. If we're successful we detect technology, from which we infer a technologist, who may or may not still be around.

WW: Couldn't you have a natural technology sending an intelligent signal, like an engineered pulsar or something?

JT: I would think that someone created the technology to engineer the pulsar. There's this great pulse that was just detected, published in the fall of 2007. A single pulse that lasted less than five milliseconds, and which was extraordinarily dispersed in frequency. And if you take that dispersion to be due to the scattering of electrons in the intervening medium, then you say, "Oh, that dispersion measure is so huge, that signal is coming from outside the galaxy." And then you say, "Well, if it's coming from outside the galaxy, it's unbelievably strong." It's got to be two black holes colliding or something, to produce that much energy. And that's probably what it's going to turn out to be.

WW: How long was the pulse?

JT: Less than five milliseconds. And there are no other pulses within the record of time the data were taken. We've now tried to look for more, because if you find one in ninety-six hours of data, there should be a lot more where you found them.

WW: Was it the galactic plane?

JT: No, it was actually in the direction of the Magellanic Clouds.

WW: On the other side of our galaxy?

JT: In that direction. The inference that it's outside of our galaxy comes from saying that dispersion is due to interstellar scattering. But it could be an engineered signal. And if it is, the technology that made it is probably not nature. My bet would be that it was a technologist somewhere.

WW: Just sending out one little blast.

JT: For whatever reason.

WW: Is that because the physical explanations are problematic?

JT: No. Because it's in the realm of possibility. And like Jocelyn Bell with the pulsars, when we come up with anomalies, we ought not to totally ignore them. If you can't say that it's black holes colliding or some other phenomenon, then let's go back to thinking about some technologist somewhere who figured out how to do this.

WW: So you must have a catalog of these signals that could potentially be engineered.

JT: Yes, there are a few of them. This one is the most spectacular. ∞

Nineteen

Complex Networks, Feedback Loops, and the Cities of the Future

Carlo Ratti and Steven Strogatz

The architect and the mathematician discuss the laws that govern urban behavior and how those laws might shape the cities of the future.

Steven Strogatz mathematically describes how natural and sociocultural complexity resolves into vast webs of order. Carlo Ratti uses technology as a tool to create interactive urban environments. Strogatz holds that we have a poor understanding of the complexity underlying such networked systems. Ratti suggests that building them may actually further our understanding. *Seed* invited them to consider: Are there laws that govern urban behavior? How do feedback loops behave in dynamic systems? What will cities of the future look like?

Italian architect and designer Carlo Ratti is interested in the dynamic between technology and physical space. His work, exhibited at MoMA and the Venice Biennale, among other venues, uses the flow of information to map and predict urban behavior and has garnered him a place on the Italian Design Council. In 2008, his pavilion at the World Expo was named one of *Time*'s best inventions of the year and he was named one of *Esquire*'s "best and brightest" of the year. Ratti teaches at MIT, where he directs the SENSEable City Laboratory, an interdisciplinary group in the Department of Urban Studies and Planning.

Mathematician Steven Strogatz has pioneered the study of synchrony and the dynamic behavior of complex networks. His research explores the common ground between such phenomena as human circadian rhythms, firefly synchronicity, and small-world networks in nature and society. The recipient of numerous teaching awards, Strogatz is author of *Sync: The Emerging Science of Spontaneous Order* (2003) and *The Calculus of Friendship* (2009). He is currently Jacob Gould Schurman Professor of Applied Mathematics at Cornell University.

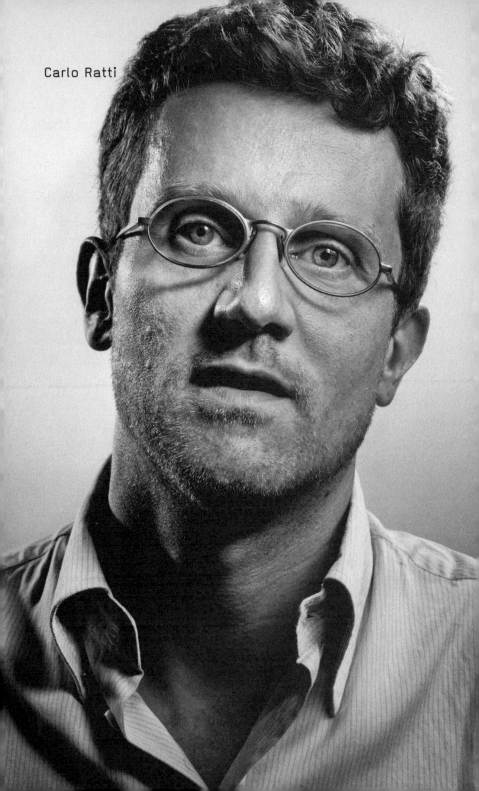
Carlo Ratti

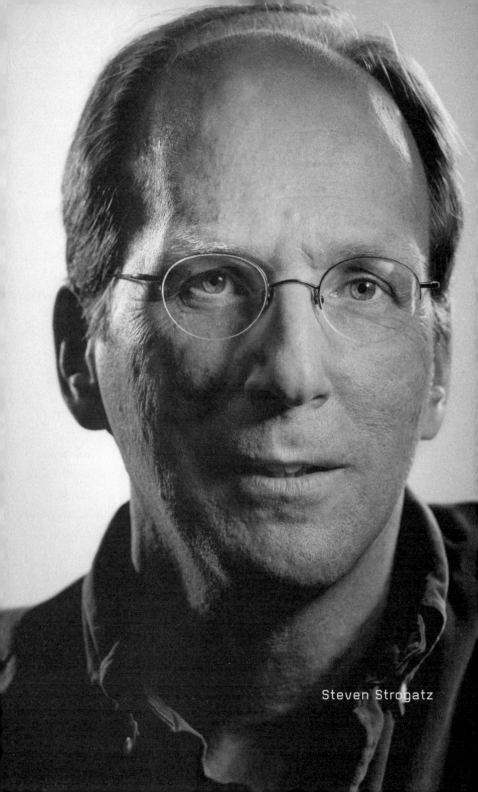

Steven Strogatz

Steven Strogatz: Tell me about the New York Talk Exchange—what were you trying to do with that project?

Carlo Ratti: Well, let me first tell you about the framework we're trying to develop in the SENSEable City Lab at MIT. We think there's a big change under way related to our physical space, our environment. We know how cities of the past were built with concrete, brick, glass, and steel. But cities of the future will also be made of silicon. So, how to marry concrete and silicon? This is what we're looking at.

In the case of the New York Talk Exchange we were trying to see how people's use of technology can allow us to better understand the city. So, using AT&T data, we looked at how New York connects in real time with the rest of the world, which gives you the effect of time zones, the effect of the specific neighborhood. For example, the rich parts of the city talk to the rich parts of the globe and so on. We wanted to know how the city could be seen differently by looking at all these connections.

SS: What kind of resolution did you use in the case of New York? Did you have it by borough or by street or . . . ?

CR: Actually, we had maps of different scales. On one, New York was just one point. But then zooming in, we could go to the 200 x 200–meter resolution. So quite fine-grained. For example, looking at Flushing, Queens, we could see that we had a huge number of calls to Guyana. And we said, well what's going on, maybe we didn't properly analyze the data from AT&T. And then we went on site and found out that it's actually one of the main Guyanese communities outside Guyana. A lot of interesting things like this came out of the data.

SS: That's actually something I wanted to get into a little bit. This process of visualizing the connections between different communities—in this case worldwide—gives us a different feel for the world we're living in. There's something kind of mystifying about complex networks and globalization in general. So many of the things that we face today, whether in ecosystems or in global climate change or economics, involve these vast networks that we really have trouble thinking about, partly because we can't see them. I wonder if these studies might give us some kind of new intuition.

CR: I think the interesting thing is that the amount of information we can visualize is actually increasing. We keep overlaying more and more previously invisible information onto our physical space. And by making it visible, by analyzing it, you can actually have an effect on people's perception of it. By looking at cell-phone networks we can see a city as a living organism, a pulsating entity.

SS: Yeah, it's very provocative to me, this idea of looking at cities this way. We tend to think of them as geographical regions filled with people, but we don't have a very clear dynamic sense of what's happening, of the pulse, as you say.

In a lot of your work it seems to me as if there's an inherent optimism that human beings interacting with each other in new ways, and with their surroundings, will generally be a good thing. Do you ever worry about this?

CR: The fact that we have the potential to communicate with more people in a variety of ways seems to me to be very, very good. It's like a democracy of opportunities. Also, from an evolutionary point of view, it's the opposite of inbreeding, being able to reach out to anybody else on the planet.

SS: Yes, that's clearly good. What I'm a bit worried about though is when people interact with their environment and the environment can act back. I don't know anything about architecture or urban planning, but I get the impression that the old view is that structures exist and then people move around them or use them, but they don't actually alter the space, by and large. But in some of your work it's clear that the structures are responding to the people.

CR: You're right. Traditionally, there was very little interaction. Architects have big egos, so they felt that the interaction would go mostly from the built environment to the people. For instance, fifty years ago Buckminster Fuller said, "Reform the environment, stop trying to reform the people. They will reform themselves if the environment is right." What's happening today, I think, is that we're able to create a full feedback loop. The two things can have an interplay, thanks to the new technology. For example, we created the Digital Water Pavilion for the Expo 2008 in Spain. The Expo's theme was water, so the idea was to make a building completely out of water. Bill Mitchell at MIT had the idea to treat water like little valves you could open and close, creating pixels made of water. So it was like a living building. When you approach it, it opens to let you in, it reacts to your presence. It becomes a fully engaging experience between the human and the building. So we're able to actuate the city in different ways than we could in the past. Why do you see this as troubling?

SS: What I'm worried about is exactly what you put your finger on: feedback loops. In the world of dynamical systems, from a mathematical standpoint, feedback loops, especially in complex systems, can be really scary. Because of their unintended consequences. They can create all the beauty and richness in the world around us as well as unforeseen horrors. Just to take a supersimple example of what I'm thinking of here, look at the Millennium Bridge in London: one of the world's thinnest foot bridges and a very elegant structure. All the architects agreed that it was gorgeous, but it looked like it wanted to vibrate, like it was practically a guitar string strung across the Thames River. And on opening day when people walked across the bridge it wobbled a little bit. Which then fed to the people and made them tend to synchronize their footfalls with the bridge's motion, which made the bridge's motion worse. None of this was supposed to happen. This was not built in.

CR: It was very funny—I was in London that day, at the opening.

ss: Oh, so you should tell me!

CR: Yeah, for a couple of days before they closed it, it actually became the thing to do in London, to walk on the wobbly bridge. But, if I get your point correctly, I think you're saying feedback loops can create instability and resonance or some type of divergence in the equation.

ss: That's one possible unforeseen consequence, yes. And, true, those would probably be easy enough to engineer against. So I'm not losing sleep over it, but I have this general concern about entering this networked era, which we're clearly already in. For example, the power grid used to not be a grid. It was just a lot of isolated power stations. When there was trouble people would just close down the power plants and repair whatever the problem was. But now that there's a grid, when something bad happens at one point in the grid, and you use the defense strategy of just shutting down that plant, it can have propagating effects. It can put too much load on other plants, which may cause them to shut down. And this is exactly what we saw here in the northeast when we had the 2003 blackout. Or think about what is happening right now in the market, where there are all kinds of propagating, cascading failures in our market and financial systems. So I'm just thinking that you may be opening a Pandora's box for architecture when you let people feedback on the built environment. There will possibly be a lot of wonderful, emergent things, but there may be some very disturbing things.

CR: I like your example of the power grid. But isn't part of the problem that we're treating new distributed systems with old, centralized controllers? What if the intelligence were spread into the system? Then the system would regulate itself in a different way.

ss: Right. That's actually the solution that nature tries to use.

So the thing is, if you're going to create these big, complex networks, you'd better be prepared to control them in a decentralized way also. The two should go hand in hand. In the case of the financial system, we don't have anything like that really. We use the language that comes from the power-grid scenario. We talk about pulling the circuit breaker when the market has a problem. It's a very crude kind of response.

CR: I see your point. My view is that the two will actually develop more or less at the same time. You cannot really develop the architecture and the control system until you have a network with which you can experiment. The two would likely develop in tandem. The other good thing about cities is that they're under so much concrete. They're very stable.

SS: Good point.

CR: Also, the city has such inertia. It's not as if when you have a traffic jam you can just make the street double the width and let it go. You can do a little bit, you can change the traffic lights, synchronize them, but you can't do much. If you get this information back to the humans, though, the humans can become the system actuators. The beautiful thing about that is if humans contribute to the intelligence of the system then it is, as a result, decentralized, intelligent control.

SS: You are optimistic!

SS: In science, we try to discover natural laws within biological or physical phenomena. With all that we're discussing, it seems like all kinds of laws for urban phenomena are waiting to be discovered.

CR: Absolutely. One of the most fascinating questions to me is the law of the scale of the city. Cities evolved for a large number of reasons. We didn't have cities eight thousand years ago and that's a pretty short time ago. They evolved for commerce, trading, defense, the industrial revolution, and so on. Today it seems that cities are mostly for helping us connect with each other, to meet and physically connect and to exchange and trade information. It's what makes cities exciting. Ten or fifteen years ago, at the beginning

Carlo Ratti and Steven Strogatz

of the Internet revolution, people were writing about the death of cities. The fact that we had a big, horizontal network and could connect every place in the world meant that cities would disappear. You can find a lot of literature in the mid-nineties about this. Ironically, the past fifteen years have actually seen the most extreme urbanization process in history. This year, for the first time, more than half of the world's population is living in cities. This has something to do with a fundamental law that hasn't yet been found. This is where we would need help.

SS: You want to take a guess? What's your guess at what this law looks like?

CR: It's related to the way we communicate as humans. From an evolutionary point of view, it seems advantageous to have a higher chance of meeting, and potentially mating, with more people, of being more connected with the world physically and digitally through networks. So what next? We've seen cities in the past decades reaching 20 million. Are we soon going to see a city reach 100 million? Why don't we then live on the planet in a way that resembles New York and Central Park: a city of 6 billion surrounded by green space? These are important questions that are deeply connected to the way we communicate with each other and structure ourselves in society.

But, the thing is, it looks like technology is changing the way we communicate and because of these new connection patterns, we need a new type of physical structure for cities. Let me tell you something interesting that we found. We haven't solved it yet, but we are working on it.

We were looking at the data generated by the New York Talk Exchange project and how New York City communicates with all the cities in the United States. Now, if you take all the calls from New York City to all these other cities, you get a very surprising power law. If you look at a city of one hundred thousand and a city of one million, the city of one million will have a hundred times more connections with New York than the city of one hundred thousand. If you think about the uniform system, you would expect a simple linear or gravitational model. You have two populations, and connections should be proportional to population. What we find instead is that connections are proportional to population squared.

SS: That's interesting. Did you happen to know, offhand, how far that extended? Was it down to cities of size one hundred or one thousand or ten thousand?

CR: It's consistent between one thousand and several million. It looks like an extremely consistent pattern. We're still working on the data, but one interesting finding it suggests is that if you're in a big city you're much more connected to the rest of the population. The old experiment of six degrees of separation, which was very interesting, but quite approximate, could be validated today on a very large scale. People have started to do it with MSN actually.

SS: Oh yes, I heard about that—Jure Leskovec's work. Interesting stuff.

CR: Yeah, they published a paper last summer. The problem is that they don't have the precise location of the people. If you did the same thing on a telephone network, then you'd know the location of the network nodes. For instance, I think it would be incredible to find out that in New York City people have two fewer degrees of separation from the world than in Iowa City!

SS: I study nonlinear systems where the causes are not just proportional to the effects that they produce, or the sum of the parts doesn't necessarily equal the whole. In these kinds of systems, which have been studied now for about a hundred years, we know that all kinds of weird things can happen that are hard to predict. Chaos is probably the most famous phenomenon—a little disturbance gives rise to the proverbial butterfly flapping its wings that starts a hurricane in Brazil. Even chaos theory, which was developed and came to full flower in the 1980s, has only really helped us understand

Carlo Ratti and Steven Strogatz

nonlinear systems with very few parts, only two or three degrees of freedom. So when I hear about all the good things that are going to happen when populations of 10 million interact with their built environment, I worry because we're nowhere close to understanding these kind of phenomena from a theoretical point of view. This is why we have so much trouble in genomics, for example, where we still don't understand why a person is more complicated than a pea, or why its organization is immensely less complicated than that of a person.

CR: I'm playing the optimist now. But the fact that we're not able to model a system into solving the equations behind it doesn't mean that the system itself is dangerous. In a sense, what's exciting about the world is that many systems seem to have been engineered with a huge safety factor built in.

SS: Those systems that have already evolved, you mean?

CR: Yes.

SS: Well, those may have evolved to be robust and have safety factors, but as we create new ones, meanwhile, we're still using our old linear thinking; that is, we didn't evolve in the world of the Web, and a world where we can affect the ecosystem or the climate. We've been doing this for only the past fifty or one hundred years, and we don't have the cortex to know what the hell we're doing. It's not even clear that we have the cortex to ever know what we're doing because these problems are so high-dimensional, and have so many parts, that even if the answer were known I'm not sure we'd understand it. So, yes, at the level of professional people working in this field, there's a lot for us to do and we'll be employed. But we can really screw things up by introducing nonlinear feedbacks into these complex networks that we're creating.

CR: But, in the end, aren't most of the systems that nature evolved also nonlinear systems? We sometimes pretend they're linear, because we know how to cope with them and they're easier to study.

But aren't all of them very complex, nonlinear systems?

SS: Absolutely. The linear world would be very gray, boring, terrible. The miracle of life and the richness of the world around us are due to its nonlinear and interconnected nature. I totally take that point, but it's taken many millions of years to get like this.

Einstein said something like, "Everything has changed except our ways of thinking." I hope that we, as a species, will get to be as fully capable of dealing with interconnectedness as the interconnected things we are creating. Can we keep up with our own creations?

CR: But isn't that how evolution works? By introducing something entirely new into the evolutionary chain?

SS: Right. And a lot of things go extinct. Most things that have ever been born on Earth, most species are done, they had their evolutionary experiments and they lost, right?

CR: Well, they mutated into something else. I mean, we are the living legacy of that.

SS: Right, they may have had survivors that were better adapted. And so, yes, if that is some consolation for us, that it will give rise to something else that will survive us, then okay.

CR: But I think some of the things we are experimenting with are really following some evolutionary principles and, I think, making us more diverse. There are a number of studies on species that went extinct and most of them seem to refer to the fact that those who had become too specialized disappeared. A small change in the environment killed them. And aren't the things that we're discussing making us less specialized? Isn't it all going in the direction of making us evolutionarily more effective?

SS: Maybe. I hope so. Generally that would be a good strategy to be more diversified in uncertain times. But I think our best way of coping with it is to try to develop new ways of having intuition about the world. With the rise of the World Wide Web, I think a lot of us started to have an understanding of what a network is, in a way that we didn't have twenty years ago. And even the language—when people speak about surfing the Web, that's a very physical metaphor. To me that shows progress in our way of thinking, in our understanding of this gigantic, seemingly endless web of connections.

CR: Speaking from that point of view, what are the key challenges from a mathematical point of view in this field?

SS: The challenges are enormous. We can't solve the equations of any of these systems that we are talking about in the traditional sense that we could solve Newton's laws for mechanical systems, for example. So the math that we all grew up on, calculus and differential equations, is still very valuable, but it's not clear that it will really suffice because we can't solve the equations. Not in the sense of getting explicit formulas for the answers. We need some other kind of insight.

CR: Right.

SS: At least in chaos and traditional nonlinear dynamics we had pictures. We could understand what was happening by visualizing these pictures in two or three, and sometimes four or five, dimensions. But now that we're dealing with network systems with millions of nodes, we can't use geometry either. We need something else. We have graph theory, that's something. We have simulation but the simulations are often as hard to understand as the reality. So I think we have a real psychological question here as to whether we will ever be able to develop what's needed to understand this modern world. And it may be, speaking of evolution, that we're not the ones who will understand it. The artificially intelligent devices that we create will understand it and they will report back to us. I think that's probably the most likely scenario. If we survive this networked world it will be because entities much more intelligent than us will figure out the laws. Who knows? The history of science has always kept moving forward, but, on the other hand, that doesn't mean it always will.

CR: Okay, so the topological instruments we have are not sufficient to analyze these huge networks. But some networks do end up in physical space, like, for example, the network of people talking cell phone to cell phone. And when you go back to the physical space, we have a huge number of traditional tools that we can use. And so I wonder if, for some networks, going back to the Euclidean space can actually help us to better analyze what's going on. We're really not sure how to structure the network, but we have already structured the physical Euclidean space that is imposing on it.

SS: Right. That's a very interesting point. Very little in network theory has been done, believe it or not, on networks in geographical Euclidean space. But yeah, on top of the pure topology of who is connected to whom is the structure of actual distances. So people have started to think about spatial networks and I think that could be a really important area for the development of network theory in the next few years.

CR: So do you think architecture could save mathematics?

SS: Yeah, sure. Lead the way! ∞

Carlo Ratti and Steven Strogatz

Twenty

Social Networks
Albert-László Barabási and James Fowler

The physicist and the political scientist discuss contagion and the Obama campaign, debate the natural selection of robustness, and ask whether society is turning inward.

Using the lingua franca of math, Albert-László Barabási describes networks in the World Wide Web, the Internet, the human body, and society at large. James Fowler seeks to identify the social and biological links that define us as humans. But whereas Barabási sees similarity across systems, Fowler believes that the underlying principle in social networks may be inherent variation. Over lunch in Boston, they discussed contagion and the Obama campaign, debated the natural selection of robustness, and asked, Is society turning inward?

Best known for developing the concept of "scale-free" networks, Hungarian physicist Albert-László Barabási is a major contributor to the field of network science. He is currently exploring the interplay between networks and human dynamics and the role of cellular networks in human diseases. The recipient of numerous prizes in biology, physics, and computer science, Barabási directs the Center for Network Science at Northeastern University and is a member of the Center of Cancer Systems Biology at the Dana Farber Cancer Institute, Harvard University. He is the author of *Linked: How Everything Is Connected to Everything Else and What It Means for Business, Science, and Everyday Life*. His newest book is *Bursts: The Hidden Pattern Behind Everything We Do*.

James Fowler is a new kind of political scientist. Specializing in social networks, the evolution of cooperation, and genopolitics (the genetic basis of political behavior), he melds the social with the biological, pushing the boundaries of his field to discover, for example, that smoking, obesity, and happiness spread within social networks, and that genes affect voting behavior. A professor at the University of California, San Diego, Fowler is famous among students for publishing the first scientific evidence to support the "Colbert bump," by which anyone who appears on *The Colbert Report* experiences a significant spike in popularity. He is co-author of *Connected: The Surprising Power of Our Social Networks and How They Shape Our Lives*.

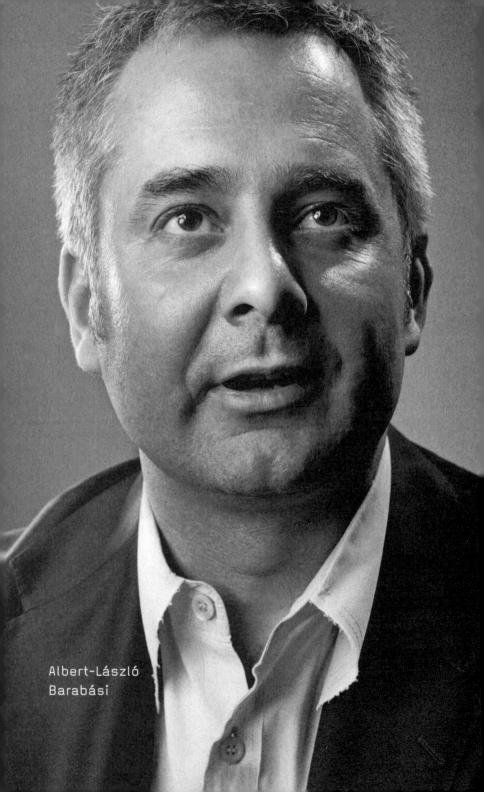

Albert-László
Barabási

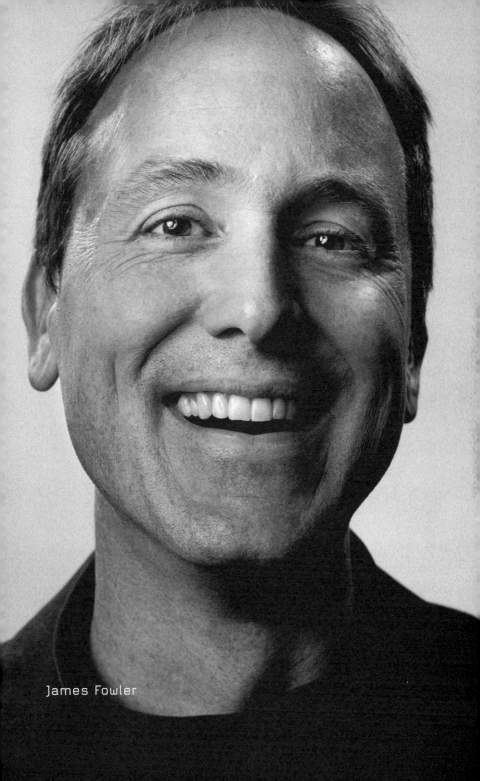

James Fowler

Albert-László Barabási: It is becoming a truism that we're living in the era of networks. Just about anywhere we turn, we encounter one. We have the World Wide Web and the Internet; we have social networks, genetic networks, and biochemical networks. These things—Web pages, genes, chemicals in our cells—are nothing new. What is new is that everybody's waking up to the fact that there is a network behind all of these systems, and we need to think about networks as a common feature of all complex systems. But I don't know if that's the way you see it.

James Fowler: Well, as a social scientist, I'm always asking, "Why do people do stuff?" So for me, what is most amazing about networks is that they completely transform the way we think about data. For a really long time, we've thought about individuals as though they were islands—a Robinson Crusoe model of social science. Being able to integrate information—not just about people, but about their *relationships*—is something that's completely new.

The rise of online social networks in the past few years has been very important in this respect. Now we can ask, "What's happening in that whole complex set of relationships that we could never learn by looking at just each individual?"

ALB: Social networks have also given us a new cache of hard data so we're no longer talking so abstractly about networks. One of the fundamental surprises, which certainly excites the physics community, is that we keep finding similar organizing principles across widely different systems. That is, if for a moment you forget that one node is a metabolite, the other is a gene, and the third is a person, the networks behind metabolism, genetics, and social systems are very much alike. And this has allowed people—social scientists like you, physicists like me, as well as biologists and economists—to talk together on equal terms.

JF: It's really breaking down barriers. And I completely agree that it's data driven, that we have this new information, especially about

Albert-László Barabási and James Fowler

interactions between people and interactions at the cellular level, that has driven an interest in these methods. But the methods are now driving interest in the data. There's a lot more interest, for example, in how to define the nodes and the relationships in, say, a set of college students in a dorm. So if we asked them, "Who are your friends?" we can now follow who they call by tracking their cellphone usage, or by tagging their phones with GPS to see who they physically spend time with.

Getting that kind of massive, passive data has now become a goal, because we have these network methods that you and your colleagues have developed.

ALB: Which means that there is this huge train of digital fingerprint data coming toward social science, capturing in details just about anything humans do. Are you prepared for that?

JF: Well, the great thing about these massive, passive data sets is that we're going to have really deep information about a very, very large number of people. So we won't be forced anymore to make trade-offs between depth and breadth. But then the question becomes: What kind of preparation are we going to give our students? We've had a revolution in game theory in the past thirty years, so that a good number of political scientists all across the country work only on mathematical, closed-form models. We've also had a revolution in the application of statistics.

But both of these revolutions have been built on this atomistic view of human beings. Statisticians make the assumption that all the observations are independent in order to be able to calculate statistical significance. Game theorists make it because, as you know, getting anything to work out in a closed-form model is nearly impossible if you assume that people are taking into account the preferences of other people.

We need not only to ramp up the amount of methodological training that people in social sciences have, but also to shift their perception into realizing that the relationships between people are important.

ALB: But that process and the accompanying change of perspective can create a huge amount of stress. It certainly did in genetics, where careers are traditionally made by discovering one gene and focusing deeply, in a reductionist way, on what that particular gene does.

Suddenly, a new generation of physicists, biologists, mathematicians, bioinformaticians, systems biologists—whatever you want to call them—comes along and says, "You know, I don't have a favorite gene. I want to look at all of them simultaneously." That is a fundamental change in the view of what really matters in biology, one that not everybody is ready for. It's creating stress. Do you see this happening in sociology?

JF: Well, in political science we basically try to answer two fundamental questions: "How do we organize ourselves to achieve something that we couldn't as individuals?" and "Once we achieve that, how do we decide who gets what?" We've made progress in terms of figuring out individual decision making, but I don't know how you get further on these questions without networks.

So for me, networks present a tremendous opportunity. It's like when Leeuwenhoek first looked at the structure of the cell and for the very first time was able to connect things inside the cell to the way the cell functioned.

ALB: Let's stay with political science for a second. The question that everybody asks about networks is, "So what? Do we ever see their consequences?" Recently people have pointed to the Obama campaign and said, "This is where network science leads you."

JF: Yes. A lot of this stuff started with the Dean campaign, realizing the importance of using the Internet to mobilize people and accessing social networks. So, rather than knocking on someone's door or cold-calling them, you help them have coffee with

Albert-László Barabási and James Fowler

one another, where they'll discuss the candidates. So you're promoting social interaction. What Obama did is take this to a larger scale.

The second thing they did, which was counterintuitive to some economists, is they allowed people to donate any amount of money they wanted to—even donations as small as a dollar. The fixed cost of them doing that was actually pretty high. But they realized that once you give a little bit, you're probably more motivated to give a lot later on. They also understood that once you give, the behavior spreads in your social networks. You say to your friends, "I gave money to the Obama campaign." They hear that, and then they're more inclined to give.

So his fund-raising was just off the charts. I'm sure that as we read the tea leaves, we're going to find that social networks were a big part of the story. I also think, going forward, that every campaign is going to be run this way, because they're going to see how successful this one was.

Campaigns in politics are sort of like evolution, right? Because only the survivor wins, and people will try to copy the better strategy.

ALB: Which actually leads to an interesting question: the spread of networks. You just mentioned how the urge to give money spreads through social networks. And you and Nicholas Christakis also did that wonderful study on the impact of social networks on health. My group studies how an error in the network within the cells spreads along the genetic links, leading to multiple diseases.

There is really a paradigm change here. We always hear about diseased genes, but what we are learning through networks is that when a disease emerges, it's typically because some part of the network breaks down in your cell. There is no single cancer gene. There are actually closer to three hundred genes associated with cancer—

JF: That we know of so far.

ALB: That's right. And they break down in different combinations, though they all lead to the same type of cancer. It's very puzzling to everybody. So I often give this analogy: If you go out in the morning to start your car and the lights don't go on, there could be lots of reasons. It is perhaps because the battery is dead or maybe a cable is broken, or the switch is not working, or your lightbulb is broken. Or maybe a fuse is gone.

Once you take it to the shop, the mechanic can simply pull out the wiring diagram and check a few points; within five minutes, he diagnoses the problem and then can replace the right component. We don't yet have the wiring diagrams of cellular networks and we are missing the spare parts, too. An important goal in biology and medicine is to get those diagrams.

So this is one paradigm change. On top of this you and Nicholas have been working on how the social network would have just as much impact on disease as the cellular network.

JF: Yes. I was interested in how political behavior could spread through social networks. So some of the very first work I did was on the question: "If I vote, how does it affect my friends and family?" Nicholas had a very similar history, but he was interested in health. He had all this work on spouses, how if one spouse dies it can cause the other spouse to die prematurely, for example. And so we wondered, "Why would it stop there?" If something is happening to me, and it has an effect on you, well, you're going to have friends, and they'll have friends, too. So even though there's only a small chance that I'll have a significant impact on you, that small chance gets multiplied. Pretty soon, in the network, you're talking about dozens, hundreds, thousands of people who are going to be indirectly influenced.

As it turns out, we found this for obesity. Weight gain and weight loss can spread from person to person to person through the network, up to three degrees of separation. We also found this for smoking and, more recently, for happiness.

We have these connections, so we can see things we've never seen before—like what happens to a person who's three degrees removed. Again, this is like Leeuwenhoek looking inside the cell for the first time.

JF: I've noticed something interesting on Facebook. I'll have a cluster of friends who are not on Facebook and when one becomes my friend, all of a sudden—in a matter of days—they will all become friends with me, and all become friends with one another, until the community is linked.

ALB: That's right. And then it freezes until some other friends come along and connect another community to you. The phenomenon is not unique to Facebook; the Web also evolves through bursts.

JF: Exactly. And in Facebook it's not so easy for links to go away.

ALB: It's a wonderful example of how our world has changed thanks to technology. We all have friends whom we've accumulated over twenty, thirty, forty years, depending on our age. I moved from Transylvania to Hungary and from Hungary to the United States. In the past, most of my previous links were lost. Now they're all on Facebook and the Hungarian equivalent, called iWiW. Suddenly, these social networking sites become a depository of our personal history. Some elementary school friends recently reconnected with me. People about whom I have to think hard—"Who are they?" Then, I remember, "Oh yes, he was in my fourth grade class." Now he's a friend.

Because of technology we have stopped hemorrhaging links. And I think this is fundamentally changing how we behave on a daily basis.

JF: I agree. But if we move from five friends in real life to five hundred on Facebook, it's not the case that we are having a close, deep relationship with each of those five hundred friends.

ALB: Sure.

JF: In fact, one of the intriguing things I've noticed about these online networks is that they have a property that's different from real-world social networks. As you know, in the real world, popular people tend to be friends with popular people. But in these technological networks, as in metabolic networks, it's just the opposite. The nodes with many, many links will tend to be linked to nodes with few links.

ALB: Right.

JF: It makes me wonder if the dynamics of online social networks are going to be reflective of real-world social networks. Because to a large extent, in your work and some of the work that I've done, we're relying on the idea that what we see online is telling us something about the real world. But there's a pretty fundamental difference.

ALB: Which brings up a good point: What do we mean when we say that all the "real-world" networks—the technological, social, and metabolic ones—are similar to each other? They share a few fundamental organizing principles. One that has gotten lots of attention in the scientific community is the existence of the hubs. All of these extremely disparate networks, from the cell that has developed over 4 billion years to the World Wide Web with a twenty-year history, have naturally developed these hubs. Somehow, networks always converge to the same underlying scale-free structure.

JF: This really takes us back to Darwin. Which for me, in the social sciences, is a little controversial. But I believe we're going to find that natural selection is what causes hubs to emerge in all these different networks. You have natural selection operating in the cell. You have it operating on the evolution of the brain. And recently, Nicholas Christakis, Christopher Dawes, and I have found evidence that there's a genetic basis for human social networks—that the number of people who name you as a friend is actually heritable, and about half of the variation in the number of friends can be explained due to variation in the genes.

ALB: So you mean my genes affect how many people would name me as a friend?

JF: Yes.

ALB: Can I get that gene?

JF: Some of this makes sense, right? Physically attractive people, people who communicate well, people who have assets, are probably going to be more attractive. But we were startled at how strong the genetic effect was. To me, what this really says is that human social networks have been operating under natural selection for a very long time—since we were walking around on the plains of the Serengeti in the Pleistocene. These forces are still with us today. So I really appreciate the effort to explain the variation between the hubs and the isolates in these physics models using a single underlying principle. But what this suggests to me is that it's not necessarily inherent homogeneity or similarity that brings about some of this variation; it's actually inherent variation.

There are things that make each of us, as human beings, unique—that give each of us a unique place in the social network. The fact that we're finding this sort of genetic relationship makes me wonder if there's actually a genetic purpose. That is, natural selection might have acted on us to make sure that we have a variety of people who are hubs and who aren't hubs, that we have people within these dense networks, as well as people who act as bridges between groups.

ALB: Wow. I have never heard that one. It's interesting, because the question of the role of natural selection came up very sharply when we first started to look at cellular networks, when we really didn't know what to expect. When the data came out, in each case we saw the same scale-free structure, which forced us to say, "Why so?"

And now the understanding is that it is because of growth—the fact that each network emerges through the gradual addition of new nodes. The growth process imposes such strong constraints on the network structure that all natural selection does is choose among the many possible scale-free networks. In the case of biological systems, we understand why the cell is scale free. What biologists have shown is that if the main mechanism by which you add new genes to the cell is by gene duplication—that is, you copy and recopy existing genes—then the only network that can emerge from that process is a scale-free network.

JF: So where does it come from? I mean, if they're all scale free, then that suggests that natural selection isn't the cause.

ALB: Right. And actually, one of the important properties of scale-free networks is their robustness. That is, if you start randomly removing nodes from a scale-free network, the network will not collapse. Which initially led us, and many others, to think, "Well, then the reason the cell is scale free, the reason the hubs are there, is because of this robustness. It's good for the cell, therefore natural selection has led the network to be scale free." Yet nobody has managed to produce a scale-free network that is built on the robustness principle. If you try to optimize a network to be very robust to random node removal, to breakdowns, you'll never find a scale-free one.

This suggests to us that the scale-free state of the cell, the existence of the hubs, is not because the cell has optimized itself to be resilient against mutations and other types of errors. It's really coming from the way the cell—just like the Internet—is created from the growth process, one node at a time. Since hubs happen to be a desirable property, there is no reason for natural selection to delete them.

JF: People have always been aware of their friends, but for the first time, we're becoming aware of friends of friends of friends. It's going to be interesting to see if this changes behavior. I know that as a result of our studies on obesity, and especially on happiness, my own behavior has changed.

You can think about it in one of two ways. If you think that you are tied to all these people that you don't know and have never met,

Albert-László Barabási and James Fowler

yet they are going to have an influence on you, you might just feel like, "Geez, I have no free will. I might as well just give up. I am just a piece of flotsam on the sea, floating up and down with the movements of everybody on the network."

But the other way to react is to take responsibility for all those people because they are also influenced by you. I have noticed that it has been easier for me to lose weight now, for example. And when I am walking home from the bus stop, I make sure to put on my favorite song. Because I know now that if I enter my house in a crummy mood, I'm not just going to make my son unhappy and my wife unhappy; I'm going to make my son's friend and my wife's mother unhappy. There are going to be all sorts of indirect, unintended consequences of my behavior, which makes me feel that I should take more responsibility.

I really think that the feeling of being connected is, on balance, going to be a very positive thing for our society.

ALB: Interesting. I mean, I don't know much about the psychological aspect of networks, but one of the things that I find fascinating is Milgram's iconic six degrees of separation experiment—where people were asked to pass messages to a certain person who has to be reached. When you look at the social network as a whole, you see the hubs. But when you look where people pass the messages, the hubs are missing.

Somehow, when people are asked to participate in a game, they're avoiding the hubs, even though they know that the hubs would be the most efficient way to spread the message. Basically, it's like saying, I would never pass it to my president, because I would not bother him with such a silly thing. So this is one example where we know that we have a completely different attitude toward the hubs than the nonhubs. It comes back to the psychology of how we're going to handle these links, and the way we handle them is that we differentiate between people.

JF: Right.

ALB: But that reminds me, I want to actually test an idea out on you—

JF: You've saved it for the end!

ALB: Yes, exactly! It's unquestionable that the twentieth century brought us rapid understandings of the cosmos and elementary particles. We developed quantum theory, built huge accelerators, went to the Moon. We explored everything, from the very small to the huge. In contrast, the twenty-first century has been called the century of networks, of complexity. And as network thinking emerges and explodes, we're also seeing a drop in the interest toward the traditional problems in science. I don't know whether the change is good or not, but we certainly seem to have lost the appetite to go to the small and to the big.

I mean, my son doesn't want to be an astronaut any longer. I've asked him many times: "Would you like to go to the Moon?" And he says: "No. I don't care about that." But he cares deeply about Facebook and about the Internet. He cares deeply about the Web. At the same time, students who in the past would have gone to physics and math now are enrolling in computer science and biology, or are trying to understand networks and complexity. So I think that this network explosion coincides with humanity turning inward.

I wondered if you felt the same?

JF: Yeah. Part of this turning inward is a function of the fact that people like you would be out of a job if you didn't start thinking about social science, right? So many of the outstanding questions in physics have been resolved that a lot of your colleagues are turning to other places to use their tools. The other part, I think, is that because we've already maxed out on the negative end of technology—with the creation of things like global warming and nuclear weapons—there has been a realization that maybe we ought to be putting our best and brightest minds at work on this question of how we all get along.

Albert-László Barabási and James Fowler

And it couldn't have come a moment too soon. Because the challenges we are going to face this century are truly astounding. It's an open question whether or not we are going to make it until the end of the century. But I think that if we are going to make it, it's only because we're able to understand ourselves better by using this new technology. That's really going to be what helps us find solutions to these problems that we face in the century of networks.

ALB: Amen. ∞

Twenty-One

The Physics of Infinity
Paul Steinhardt and Peter Galison

The physicist and the historian discuss the cyclic universe, the limits of physics, and the challenge of infinity.

Paul Steinhardt's "cyclic model," a radical alternative to the big bang and inflationary cosmology, proposes that the universe's evolution is periodic and that key events shaping its structure occurred before the bang. Peter Galison studies historic fundamental shifts in physics and what types of evidence count as truth. Having first met during their graduate school days at Harvard, they were quick to accept *Seed*'s invitation to consider: Where is the line between physics and metaphysics? Is infinity unscientific? What is it, ultimately, that we want from science?

A theoretical physicist whose research spans problems in condensed matter physics, particle physics, astrophysics, and cosmology, Paul Steinhardt is the Albert Einstein Professor in Science at Princeton University and director of the Princeton Center for Theoretical Science. He introduced the concept of quasicrystals, a new state of solid matter, for which he won the Oliver E. Buckley Condensed Matter Prize in 2010. He is also known as one of the architects of inflationary cosmology, for which he shared the 2003 Dirac Medal, and was among the first to show evidence for dark energy and cosmic acceleration. With physicist Neil Turok, he proposed an alternative to the standard big bang theory, known as "cyclic universe," and co-authored *Endless Universe: Beyond the Big Bang*.

Peter Galison, the Pellegrino University Professor in Physics and History of Science at Harvard, focuses on questions such as What, at any given time, convinces people that an experiment is true? The recipient of a MacArthur "genius" award and the Max Planck Prize, Galison has launched several projects examining cross-currents between science and other fields, including two documentaries: *Ultimate Weapon: The H-Bomb Dilemma* and *Secrecy*, about the costs and benefits of government secrecy, which debuted at the 2008 Sundance Film Festival.

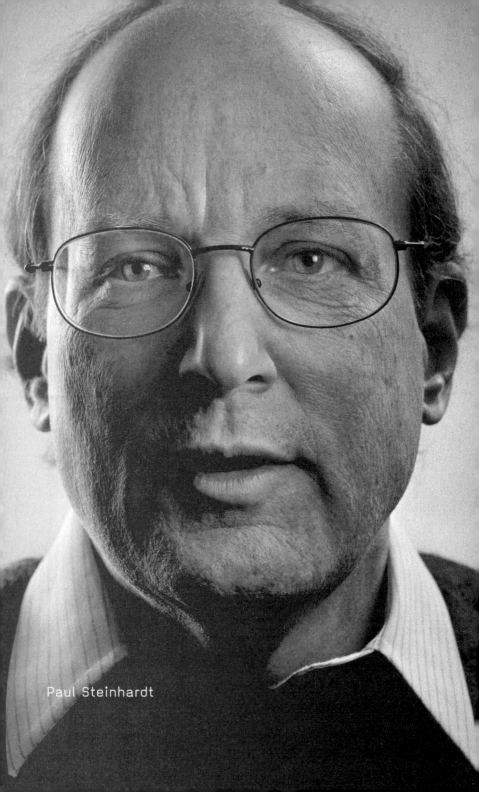
Paul Steinhardt

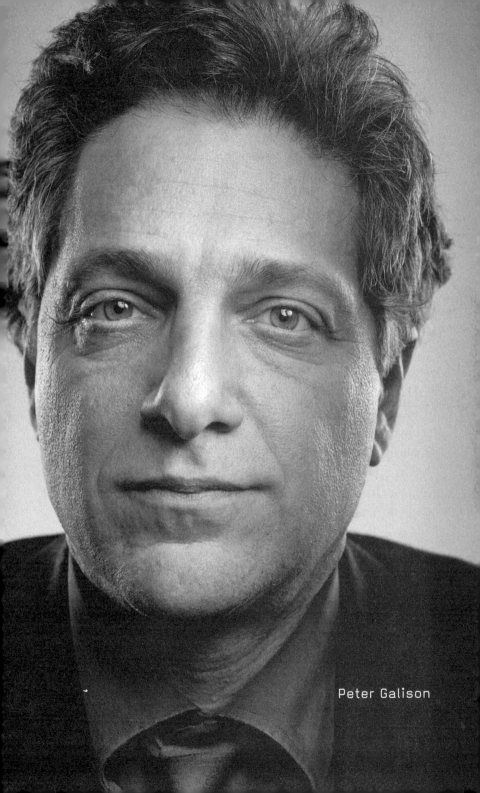

Peter Galison

Peter Galison: So, Paul, you've been thinking about the big bang and its alternatives. Where do we stand now with that?

Paul Steinhardt: Well, in cosmology these days, we have two competing ideas—ideas that lead us down two different paths for how the universe came to be, how it evolved, and what will happen in the future.

The standard view is that the bang is the beginning, that energy, matter, space, and time all came into existence at that moment. But if that's the case, then we have to describe everything we see in the universe as having taken place within just 14 billion years. Furthermore, since we know that the conditions at the one-second mark were quite special, we need something remarkable to have happened within that one second.

This has led to this idea that the universe underwent a brief period of inflation during those first few instants after the big bang. This faster-than-light expansion would have transformed the turbulent and warped universe that erupted into the smooth, uniform universe that existed by the time the universe was one second old.

But there is an alternative idea, developed in recent years, that the big bang was not the beginning. It's a moment when a lot of matter and radiation were created, but space and time existed before, as well as after. If that's the case, then suddenly you have a lot more time and new possibilities for setting up the conditions we know had to exist one second after the bang.

PG: One of the things that strikes me, thinking about these issues historically and philosophically, is that physics has been in a long pursuit of trying to eliminate the special, the particular. Galileo, for example, said that there is no special frame of reference, such that any constantly moving ship, as it plies its way through the Mediterranean Sea, ought to have the same mechanical physics as every other ship. Then Einstein extended this beyond mechanical things: he said electricity, and magnetism, and everything should be the

Paul Steinhardt and Peter Galison

same in different frames of reference. It seems to be also much of what surrounds the current cosmological debate—how to take away specialness in some way.

I mean, part of the motivation for getting away from the big-bang picture has been the idea that too much had to be specified, too much had to be sort of set up.

PS: Right. The distribution of matter and energy in the universe had to have been extremely uniform throughout space. Space itself—which, according to general relativity, can curve, warp, wrinkle, and wrap itself up in all sorts of complicated ways—had to have been remarkably flat. Not perfectly uniform, or the universe would never have evolved stars and galaxies, but with tiny deviations of just the right sort.

All of this seems so unlikely that there's no reasonable notion of how it could happen. So that is what led to the invention of inflation—an incredible stretching that results in near-perfect flatness and uniformity.

But the picture isn't as simple as it's often portrayed, and that's where it gets interesting. The original idea—the way it's often talked about in literature and textbooks, even the way we talk to students—is that inflation makes everything in the universe the same. What we've learned is that inflation actually divides the universe up into little sectors that are all different from one another. Some regions of space would be habitable like ours, but others would be uninhabitable; still others would be habitable but would not have the same physical laws or the same distributions of matter that we see here. In fact, what we see is very likely only relevant to an infinitesimal fraction of space.

PG: In a way you could treat this multiplicity of worlds as solving one problem and creating others. In some ways you could say, okay, we have 10^{500} or $10^{1,000}$ of what exists. They have different laws—

PS: Actually, you get an *infinite* number of everything. So an infinite number of patches that look like ours, but also an infinite number that would be more curved and warped than ours. And if you add some attributes from string theory, they might have different physical laws.

Because you have an infinite number of everything, you have no rigorous mathematical or statistical way of computing a probability—it's not even a sensible question to ask. So people are in the process of trying to regulate this infinity. For example, they try to invent a rule for deciding probability that makes what we see likely. But there's no way of deciding why that rule instead of some other one. They simply keep trying until they've found the answer they wanted. Some people are going down that path and are prepared to declare victory if they find something they think works.

Others take a different path. They accept the infinity of infinities and the fact that they can't find any measure for deciding whether our circumstance is more probable or not. They'll be satisfied with the fact that at least some patches look like what we see, and I will declare victory on that basis.

Personally, I don't find either of these approaches acceptable, which is why I have developed an alternative picture in which the big bang is not the beginning. A big bang repeats at regular intervals of a trillion years or so, and the evolution of the universe is cyclic.

PG: It's interesting to me how in the cyclic model—and also in inflation and in string theory—there's this idea that you don't want to make a whole lot of choices at the beginning that are too fine. Because that seems to violate the spirit. It goes back to even the days of Descartes.

Descartes had a sort of view that nobody holds now, but that was the principle. He said, "You start with just extension and motion and that's all." There are all sorts of phenomena that he tries to explain that way, but in the end he wants to explain the visible universe with this relatively simple start.

In a way, we keep that story, that attempt to have relative inde-

Paul Steinhardt and Peter Galison

pendence of our starting assumptions to explain the specificity we have today. In a way, it's what drives these ideas of infinity in these different camps: people that want the landscape to have an infinity of craters, each one a different universe, and people that want infinity in space the way Descartes did, wherein there is no special region at the center of all things. It seems to me this same spirit lies behind what you and your colleagues want to do with the cyclic process. You say, "We don't want to tune things, we don't want to manufacture a universe that seems like it's done with forethought."

PS: Right. The fewer assumptions you have to make, the better—Occam's razor. But there's another issue, which connects to why I personally do science: science's power to explain and to predict.

We've been talking about an example in which you have a complex energy landscape and an infinite number of possibilities for the universe. But we have no real explanation as to why things are the way they are, because it could have been different.

So it has no power. And without real explanatory power, it's not interesting to me. But I'd be interested to hear how this has played out in the history of science.

PG: Well, sometimes what pulls toward prediction may not pull in the same direction as explanation. Theories that are purely predictive, that don't give us a sense of understanding, can be valuable in some moments. That kind of positivist urge was part, though not all, of what Einstein was trying to do with special relativity.

He said, "It's all very well to try to understand what's going on inside of an electron and to wonder about the dynamics of the ether and whether it moves around like little gyroscopes at each point. But this is beyond what we can know. Let's start with something that we can touch and measure. We'll say the measure of space will be laying down rulers, and the measure of time will be what I can see on my watch. And we'll build up from these founded principles."

Other times it's the explanatory force of something that drives us even where prediction is difficult. Einstein's later work with general relativity, for example. When he saw that the motion of Mercury as it went around the Sun violated everything people knew from very accurate calculations based on Newton's theory, he wrote a postcard to his mother, saying, "Something snapped inside me." He knew at that moment this theory was right. He had one measurement—an old measurement, at that.

This play between prediction and explanation is a tension that lies very, very deep inside the history of physics. There are moments when, by turns, one plays a more dominant role than the other. Of course, we'd like to have both.

PS: But at least in the history of science up until the present, there's always been the thought that we're working toward a positivist-type solution. We hope to eventually find a theory that explains what we know and that predicts new things we haven't measured.

The kinds of theories we're talking about now explain after the fact, or are designed to match what we already knew. It seems to be a different kind of paradigm as to what science is, what even constitutes a valid, an acceptable, theory.

PG: Right. So there are splits in science where people disagree about something that's predicted. But there are also splits where one side says to the other, "That's not really science. It's not the project we should be engaged in." And that's a deeper, far more unsettling moment.

We have that sort of split right now among the string theorists. One side says, "Look, what's really scientific is to say there's this infinite or very huge number of craters to be imagined in some landscape, each of which carries different physical particles and different physical laws and so on. And we happen to live in one of them."

But the other says, "You've given up! You've given up the historical project of science. We went into string theory because we wanted to produce a theory that had one parameter, or very few movable parts. And now instead of a glider you've got a helicopter with ten thousand little pieces that have to move exactly the same way. If the slightest thing goes off, it falls to the ground in a heap of burning aluminum."

It's really an interesting moment in that way.

PS: I think it's historic. There's a certain community that feels, "This is an 'aha' moment. Science has to change. We have to accept that science has limits. There's only a certain amount that we'll be able to predict. Beyond that we're going to accept that we live in some special corner of space in which seemingly universal laws— including Newton's law of gravity—are just local environmental laws that aren't really characteristic of the whole."

Other groups say, "Hold it, this is failure. We either find ways of fixing the problems in those theories, or we scrap them and replace them with something else."

PG: There was a huge debate in Germany in the nineteenth century called the Limits of Science debate. The question was, "Is there a limit to what you can explain with science?" Not just the current science, but science in general. Was it a matter of *ignoramus*, that we don't know, or *ignorabimus*, that we cannot know?

For some people, it was a protection of science against the mystical or the religious, a guard against intrusions of the nonscientific into the scientific domain. It said, "Here are the limits of science; everything within this domain is ours, and beyond it, do what you want, but it's outside of what's really going to be explained by science."

PS: You mean ever?

PG: In principle. Of course, a boundary keeps one side away from the other, and vice versa. So certain figures on the religious side said, "Oh, the limits of science are a good thing." It says that there is a domain that is inherently mystical or spiritual and science won't intrude on it.

But the assertion of the limits of science was also criticized. For some people it was outrageous that you would say there's anything outside of science. This was certainly the view of the scientists who helped form the Vienna Circle—the beginnings of the philosophy of science—after the First World War. Their view was that you could put things in logical form based on observation combined with the new formal logic. If you couldn't, you were dealing with errant nonsense.

That's such a hard and fast line that it was inadequate for many scientists, like Einstein, who said that science has to go beyond the simply observable. By the end of his life, he became increasingly convinced that reality could be ascribed to phenomena beyond what we observe.

PS: So Einstein brings up the issue of space and time, which is something I want to revisit. Because although in relativity we're used to them being on equal footing, in cosmology they really aren't.

When we say the universe is 14 billion light-years across, it's somewhat unclear exactly what we mean. But, roughly, there's a certain finite patch of space we can observe because that's the greatest distance from which we can see light. We believe that the universe extends much farther than that, and it may even extend infinitely farther.

So here's the question: Is it finite—does it eventually close on itself—or is it really infinite? All that we can tell from measurements in our own patch, especially recent measurements by the WMAP satellite, is that we *can't* tell. We'd need to be able to go way beyond what will ever be visible to be sure.

So we must ask ourselves: Is this question—whether or not space is finite—even a scientific one? If it's forever beyond our realm of testing, is it science? Is it metaphysics? Is it even an important question?

PG: Is it *in principle* beyond the realm of testing?

PS: When the universe is filled with matter and radiation, the expansion of the universe continually slows down. Because of that deceleration, we are able to see the light from greater and greater distances as time goes by. At any given time, of course, there's a horizon, a limit to how far we can see. The universe is only about 14 billion years old, so the current horizon is 14 billion light-years. But in principle, as long as the expansion rate is slowing down, we can see more and more space.

That whole view has recently been upended because we've discovered that beginning about five billion years ago, the expansion of the universe has begun to speed up again. When the universe undergoes acceleration, the opposite happens—we will actually be losing sight of the space that we're seeing today.

The billions of galaxies we see now, almost all of them beyond Andromeda, will eventually recede from view forever and our patch of space will become a wasteland.

So, whether or not there's an infinite amount of stuff out there, as both eternal inflation and the cyclic model would suggest, is therefore *not* testable as a matter of principle. Accepting this kind of limit seems like something new.

PG: It's interesting how recent cosmology really is. In some ways, of course, it's the most ancient of all speculative systems. Where does the universe come from and how did it begin? What is the structure of things?

But in its modern scientific form, as a subject that can be studied experimentally, or at least observationally, it's recent. I mean, very shortly after making general relativity, Einstein began to think about what it would mean on a cosmological level, but that's only since 1917.

PS: Yes, science is a very thin veneer on what is an ancient subject.

PG: And even more recently—the first glimmers were in the thirties, but it really developed in the 1950s—came the idea that you can actually use ordinary physics to study it. That's amazing.

PS: I agree. What's happened in the last fifty years in cosmology has been a remarkable use of atomic or nuclear particle physics to enhance our understanding of the history of the universe. But progress has been limited to exploring the period between the one-second mark and today, where we know the physics can make observations. We now understand this period in such remarkable detail that if you try to develop a story of what happened before or what will happen in the future, it's incredibly constrained by all this information.

And because we have been focusing on this intermediate period that we can measure in fine detail, most of us haven't given attention to the larger questions outside the current empirical domain, like the infinity of space or the beginnings and future of time.

But now our noses are pushed against those issues. And most of my colleagues, myself included, are not really trained in history or philosophy or the earlier thinking on these issues. We are, I feel, somewhat naïve in this respect.

PG: To me, this again broaches the deeper subject of what it is that we want from science. Is it meaningful to talk about another universe to which we have no observational access? Is it meaningful to talk about what happened before the big bang? Is there a way of taking the cyclic model into account to try to figure this out?

In many ways the ultimate paradigm shift is for us to define what kinds of questions we consider within the domain of physics, and within the domain of science in general.

PS: I view myself as a very pragmatic positivist thinker, not so much because it's a conviction, but because it's what I like to do. So in developing the cyclic model, the important question to me is *what can we explain and what can we predict* if the universe is cyclic. If it is cyclic, this bang can repeat over and over again. What happened 14 billion years ago would be the most recent bang, but there would be ones before and ones in the future.

This is the practical question: Can we prove through observation that there was something before the big bang that led up to it and produced the present universe? If so, I think we can even say something about the future, how long before the next bang occurs and what will happen in cycles to come.

But you might also ask questions like, Do we know if the bangs went infinitely far back in time, or was there a "first" cycle? I don't know yet of any empirical way to test these questions, so as long as that's the case, I'm not interested—I want something I can deduce.

And this deductive approach has led to some very interesting developments within the past decade. So, remember I said earlier that the only way we could set up the right conditions after the bang was to have this period of inflation?

PG: Yes.

PS: When you let go of the idea that the big bang was the beginning, it turns out there is a new possibility: Before the bang, there was a period of a very special high-pressure ultra-slow contraction. And, surprisingly, it has an effect very similar to inflation's very rapid, low-pressure expansion. The physics and the timing are completely different, but it has the effect of smoothing the universe out before the bang and producing tiny ripples at the bang that look almost identical to—though not exactly like—what you get from inflation.

Close enough that at the present time we can't distinguish the two, yet different enough that we are at the edge, experimentally, of saying whether or not the cyclic picture could work. There are measurements being done right now that could tell us whether or not the idea is viable.

Soon—and it could be anywhere in the range of days to years—we could learn if the cyclic model is dead. If that's the case, then you have only the possibility that the big bang is the beginning.

PG: In the long run, it seems that you have in the history of science strong explanatory structures that have to be coupled to some kind of predictive strengths in order to go forward. If you think about it, for instance, understanding the theory of gases as an assembly of molecules bouncing went on for decades with the goal of being able to reproduce what people already knew had to do with heat in the physics of thermodynamics.

PS: But that's more ordinary science. Consider the following problem we have right now in cosmology: The cyclic model predicts that there shouldn't be a spectrum of large-wavelength gravitational waves, and typical inflation says they should exist. If we see them, as far as I know, the cyclic theory is dead, and the case for inflation is strengthened. That's normal science.

But suppose we don't observe them. That would still be consistent with the cyclic picture. But in the inflationary picture, because it allows an infinite number of patches of every possibility, there will always be some patches that actually don't produce gravitational waves. Since you get an infinite amount of everything, it might be that we are living in one of those patches of the universe. How do we prove or disprove that idea?

It's a very strange kind of science. To me, the playing field becomes uneven between a theory making testable predictions versus a theory that is Teflon. From what I can tell, there is no observation that can disprove inflation. Some people even advertise that as an advantage. To me, that's a new and unacceptable way of thinking.

PG: I think we do live at a time when science is changing its character in lots of ways, some of them no doubt for the better. Fascinating new combinations between pure science, applied science, nanoscience, bioinformatics—areas that twenty, thirty years ago might have been considered industrial stuff to be done outside the university—are now an extremely dynamic, exciting part of science.

Paul Steinhardt and Peter Galison

Cosmology, it seems to me, has always been on a very delicate balance between its speculative origins that take it outside of the remit of science and tremendously exciting moments when things previously thought to be outside of our possible understanding are suddenly understood. We're at that knife edge right now. ∞

Twenty-Two

Smarter Infrastructure
Thomas E. Lovejoy and Mitchell Joachim

The biodiversity expert and the architect/urban planner discuss victory gardens, vertical farms, senators in the jungle, and the need for diverse substainable infrastructure.

Since 1965, when Thomas Lovejoy began exploring biodiversity in the Amazon, he has worked to defend ecosystems against human impact. Mitchell Joachim creates human habitats and transportation systems— and believes that ecology can inform a new, more resilient ethos of design. When the two met in Manhattan, the ideas flew: from victory gardens to vertical farms, senators in the jungle to students toting trash. And from their discussion, a critical question emerged: Are we a society of Homer Simpsons or WALL-Es?

Tropical biologist Thomas E. Lovejoy was instrumental in establishing the field of conservation biology, introducing the term "biological diversity" to the scientific community in 1980. He is currently biodiversity chair at the H. John Heinz III Center for Science, Economics, and the Environment in Washington, D.C., and previously served as the World Bank's chief biodiversity adviser, senior adviser to the president of the United Nations Foundation, and on councils under Presidents Reagan, George H. W. Bush, and Clinton. Lovejoy received the 2001 Tyler Prize for environmental achievement.

Architect and urban designer Mitchell Joachim is co-founder of the New York–based Terreform ONE, a nonprofit design collaborative whose projects focus on integrating the principles of ecology into an urban context. Formerly a member of MIT's Smart Cities group, he has been involved with numerous award-winning projects, such as Fab Tree Hab, which exhibited at MoMA, and the Compacted Car, dubbed one of *Time*'s best inventions of the year in 2007. Joachim is currently an adjunct professor at Columbia University and a visiting professor at Parsons the New School of Design.

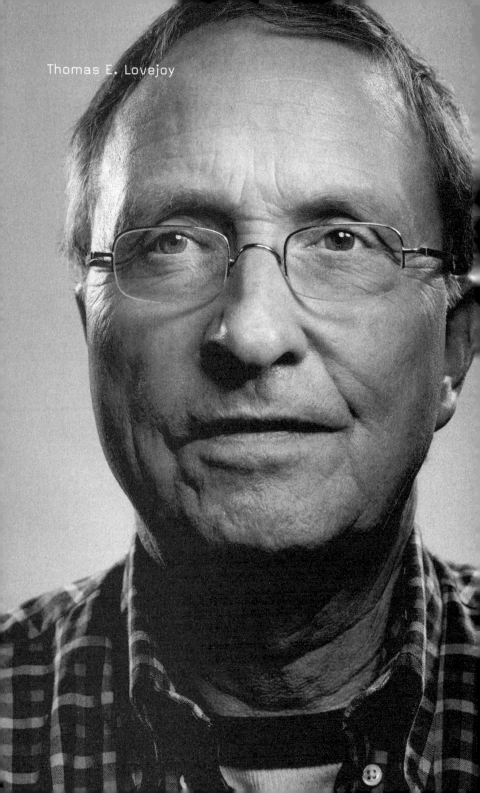

Thomas E. Lovejoy

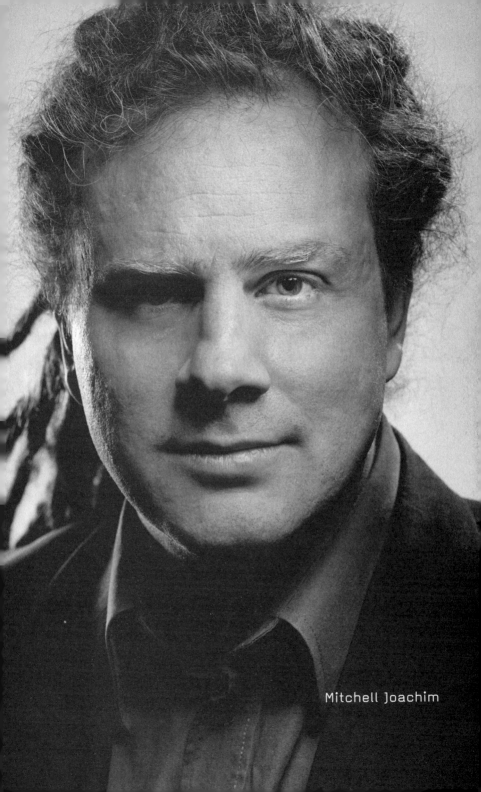

Mitchell Joachim

Mitchell Joachim: If we expect another 2.5 billion people on this Earth, and a lot of them are going to be urban dwellers, we have to rethink the concept of the city. So a lot of my work centers on that—everything from infrastructure and mobility systems to designing vehicles, cars, trains, and public housing.

Thomas Lovejoy: That's great. As I look at the global scale of our environmental problems, it's all about accommodating human aspiration in a way that's less damaging to the Earth. As Bill McDonough would say, it's so much about design.

But I'd love to hear more about some of the specific things you are doing, like "soft cars."

MJ: At MIT, we were charged with designing the car of the future. It was boring, because "of the future" quickly becomes anachronistic. There's the future car; there's been the future house. So instead we thought to design products, or a lexicon of ideas, that would fit into every car. We decided to rethink mobility, period, starting with the wheel and moving on to things like chassis and ownership and identity—bigger concepts.

"Soft car" became one meta-thought, in that if we could put the entire vehicle on these very smart wheels, with just enough power to work in the context of a city, the rest of the envelope could be freed up for anything.

It could be made of any material—pleated, soft, scuffable materials, surfaces that are okay rubbing against one another. Vehicles in this kind of system would move in a gentle congestion, in flocks and herds. Their patterns would be controlled computationally, but locally, you could switch over to user control.

TL: I've spent decades looking at development challenges in the Amazon Basin. And the worst thing that can happen there is for somebody to build a highway. Once you do that, spontaneous colonization occurs, deforestation becomes rampant, and you basically lose control.

Bruce Babbitt and I recently wrote an op-ed in Brazil's largest paper about a project almost nobody here knows of—a continent-scale infrastructure plan for South America. One of its key pieces has already been funded and built: a road from the western Amazon of Brazil, through the rainforests of Amazonian Peru, and up the Andes to Cuzco, the ancient Inca capital. It's a recipe for disaster. We basically said that the whole thing needs to be revisited in the light of current knowledge. We're not dismissing the need for transportation or the integration of South America; we're saying that we need to consider other options. The whole Amazon worked on river transport until fifty years ago. Where does the river system figure into this new plan? And where is the place for railroads? Where should cities be? How do the economics of those cities relate to the forests?

MJ: Right.

TL: My sense is that people don't care how they get somewhere as long as they can get somewhere.

MJ: Speed might be a factor at some point, but I agree. I like that you've taken a holistic, top-down approach to thinking about the overall infrastructure, at the scale of a continent. Eisenhower did this in the United States with the interstate highway system.

TL: It almost killed the rail system. There are always huge subsidies for the highways and almost none for Amtrak.

MJ: Well, it looks like the Obama administration is going to be concentrating on infrastructure for the next several years. All these things we've been talking about for the last two or three decades—decentralizing the grid, making it renewable—could probably reach fruition. It's a tremendous opportunity. So we've begun to formulate a new field that would look at the infrastructure of cities within the rubric of ecology. The idea would be to design them from scratch or to fit them into some kind of circumstance like the Amazon.

Do you know of any examples, in and around the Amazon, that fit the paradigm of a large-scale, sustainable infrastructure?

TL: The river system is one. It's vast, with 20 percent of the world's river water, and it's cheap, too: if you can get soybeans from southern Brazil up into the Amazon and send them down the river, it becomes inexpensive transport because you've got the current working for you. It's almost like a local bus service.

MJ: But the population inside the Amazon is negligible compared to a mega-city, right?

TL: At the moment, it's about thirty million. It was just three million when I first went there in the 1960s. Again, I think the key with cities is to create appealing urban centers with activities and labor that benefit the people, so that they don't find it necessary to reap short-term profit by clearing forest.

The city of Manaus, for example, is now an economic free zone, with a lot of small-scale manufacturing. This has created a nucleus of economic activity, both attracting people to Manaus and keeping the local population—roughly two million people—within the city. The state of Amazonas, of which Manaus is the capital, has the lowest deforestation rate of any of the Amazon states.

MJ: It sounds like you're trying to design around the issue of containment, trying to stop people from destroying the forest.

TL: Well, I would think about it the other way around: instead of containment, more the attraction of cities that offer a reasonable quality of life.

MJ: What do you think of things like vertical farming? You essentially have a tower, and inside you grow things like dwarf wheat and dwarf corn with robots to service it all. You recycle water and waste, all within the same system, and locate this thing in a downtown core, so that food doesn't have to travel a great distance to get to populations.

TL: So, the ultimate in local food. Has anybody done the numbers for that?

MJ: Supposedly, Dickson Despommier at Columbia University has. Are you familiar with the idea?

TL: Only very vaguely. I just wonder how you get enough solar energy.

MJ: The idea is that with the right geometry of the tower, you can get it with a lot of light wells and reflectors, or with heliostats to focus light in the building. The problem, from a policy standpoint, is that it would be very hard to put something like this in the Bronx. A tower devoted to vegetables, when so many families need public housing. So some people are working to integrate housing and food, to create urban farms within the places where people live.

TL: I've spent about twenty-five years looking at the interactions between the living planet, biological diversity, and climate change. Due to climate change, we're already seeing threshold changes in ecosystems—a basic breakdown, for instance, of the fundamental partnership that makes coral reefs. We're seeing widespread mortality in coniferous forests in North America and Europe because more pine bark beetles are surviving milder winters. Seventy percent of the trees are dead in some places.

Sitting here and looking at that—knowing that we have already increased global temperatures by three-quarters of a degree, and that there's another half coming—I get a pretty desperate feeling.

MJ: We already have done the damage.

TL: A lot of it. And so the question then becomes, how can we limit the peak amount of climate change? The answer is biology.

MJ: What do you mean?

TL: We've lost 200 to 250 billion tons of carbon to the atmosphere—not from fossil fuel usage but *in addition* to it—from deforestation, degraded grazing land, and impoverished soils. Illinois prairie soils used to contain something like 20 percent carbon; now they have 5 percent. A big chunk of that could come back. It means reforestation, restoring degraded land, and agricultural practices that build up carbon in a living soil system. If we do that, we can pull maybe

150 billion tons out of the atmosphere. It will be good for the ecosystems and therefore good for people.

MJ: You're preaching to the choir on that one. What I struggle with is my metric, which is Homer Simpson. How do we get this average American to not only comprehend the damage that's been done, but also that he needs to put this on the front of his agenda, when so many other things, like the economy, seem more important?

TL: The answer, the simplistic answer anyway, is green jobs.

MJ: Green jobs?

TL: It takes labor to do what I'm describing.

MJ: But Homer Simpsons don't like labor.

TL: Well, I think there is something here that's equivalent to the victory gardens in World War I and II. People grew these little fruit, vegetable, and herb gardens to help with the wartime public food supply. But it was also intended to boost morale, to give people a sense of empowerment through their work.

MJ: I like that example, but there's a difference in the level of adrenaline. We had world wars. We had Nazis. People were compelled because guns were facing us.

TL: We had a leader who articulated the challenge.

MJ: Yeah, we had some good leaders.

TL: And now we have one again.

MJ: We do, but the gun isn't that obvious this time. I mean, we've been saying for some time that climate change is serious, but it hasn't permeated.

TL: Well, it's really important to make it tangible.

MJ: Right. I spend a lot of time showing people doomsday scenarios—depictions of what would happen to Manhattan with a three-meter or twelve-meter rise in sea level. Still, showing someone an image isn't as powerful as a gun to your head.

TL: Yes and no. In one of my talks, I show a map of where there will be sugar maples in the future, if we continue on today's path. It's

Thomas E. Lovejoy and Mitchell Joachim

not going to be in the United States. That shocks people. There's something about autumn foliage, about sugar and syrup, that they consider part of their identity.

MJ: Certainly it's up to us, as professionals, to take the lead on what it means to "do green." I make this analogy to a razor that I own—a Gillette Fusion. The thing has five blades and it's fantastic. Now, I don't promote Gillette for any reason whatsoever; it's just the best thing for my face.

So I, the supposed Mr. Eco, am not going to use an "eco blade" because it's going to scratch up my chin. I'm going to use this poisonous thing that is packaged four times, is impossible to recycle, and takes an intense amount of embodied energy to manufacture, because it performs.

It's up to us to design the Fusion, only it's got to be cheaper and better for the environment. Consumers will choose the eco version because they like it. If it happens to be good for the Earth, that's just a side benefit.

TL: One thing you're battling against is that energy has been so cheap in this country. Not even realistically priced, but unbelievably cheap. So that most people never think about the energy in hot tap water or the difference in energy loss between a normal door and a revolving door.

MJ: Well, that's up to us, to spec those particular devices. In Paris you can't take a Hollywood shower because you have to press a button constantly to get the water flowing. This kind of design forces you to relate yourself directly to using the resource. But change is hard because Homer Simpson doesn't really want to shift his everyday agenda. Maybe I don't have enough faith in my fellow Americans, but somehow I always come back to a technocratic scenario where we have to figure out these systems behind the stage. Maybe I'm wrong. I hope I'm wrong, actually.

TL: Well, one of the things I've discovered is that if you take these mythical creatures called decision makers, and you get them away from the distractions of telephones and staff, and give them a first-hand, tangible experience, it can do wonders. So far I've been able to take twenty-one U.S. senators down to the Amazon.

We spend the night in hammocks, and let me tell you, the former vice president really snores! So did John Heinz, but he said he'd been communicating with the howler monkeys. To be fair, there's a certain level of self-selection, but still, they tell me it wasn't "real" until they saw it.

MJ: I have some assignments to get my students thinking about some of these issues we've been discussing. One is, I offer them an instant A in the class if they can confront their waste for two weeks. They carry a transparent garbage bag everywhere they go, and whatever they throw out has to end up in that bag. It averages about eight pounds a day, for New York City trash. To this day I've had zero students take on the assignment.

TL: That's so interesting—I grew up in Manhattan, in a relatively privileged household. But my parents were sufficiently marked by the Depression years that they were pretty frugal. When aluminum foil first became available, we washed and reused it until finally it fell apart. I had the job of putting the trash out, which I did not enjoy, but it meant I knew how much went out the back door.

I compare that to what goes out my back door today, and it's grown by a factor of ten. But quality of life hasn't improved by a factor of ten.

MJ: We've also got this chart of Manhattan that shows the amount of sewage, the amount of garbage, basically all human impacts on the land from the 1600s to today. Currently, the island produces waste at roughly thirty-six thousand tons per day. So we've been doing these models where you remake Manhattan, at full scale, based on the amount of garbage we've been throwing out since the time of the Indians. We can get seven full-scale models out of that. You can make up an Empire State Building in two weeks, just by the

Thomas E. Lovejoy and Mitchell Joachim

amount of paper we throw out. So we've devised these . . . You've seen the film WALL-E?

TL: No.

MJ: You haven't seen WALL-E?!

TL: No.

MJ: Oh my God. It's the best. I deeply implore you to see it.

TL: Just give me the CliffsNotes.

MJ: A cute little robot, WALL-E, is inhabiting Earth by himself because humans have left, and they've left behind nothing but trash. This robot is still on some program to pick up the trash and compact it in piles. He's a little Nebuchadnezzar; he basically makes ziggurats of trash, all day long, for no purpose except that it's his last program. He's solar-powered, so he's going to work forever.

The rest of the human race lives in some spaceship, and they're floating in chairs attached to digital devices. They're super-overweight and have no interest in other people.

TL: It seems like we're going there.

MJ: The message is pretty clear. But also the idea that the robot can form, shape, and puzzle-fit pieces of trash together is really interesting from an urban-design perspective. WALL-E was a big impetus for this project we're calling Rapid Re(f)use, which considers how to remake cities from their own waste. We've got some funky researchers working on it now. It'll look fabulous.

TL: Basically, I have a double track: One is trying to push policy at a national level or higher. The other is trying to articulate the issues for the wider public, because if the majority of the average citizenry doesn't get it, it becomes very hard to lead them.

Now the strange thing is that when you go outside of the United States, it's as if we've been living in a bubble.

MJ: So true.

TL: To be fair, it was partly that we were in a September 11 kind of coma, but it didn't have to last eight years. You go to Europe, to South America, to Asia—environmental stories have been in the press for years. People may disagree about how urgent the problems are or what to do about them, but they're aware.

MJ: Well, China is in its own bubble, too.

TL: Actually, the Chinese leadership knows it's a serious problem because all their water comes from rivers, which originate in retreating glaciers in Tibet. They know. And I think we, too, have finally taken a step in the right direction: On Secretary Clinton's first trip to China, climate change was one of the two issues she brought to the table. If we treat the Chinese with respect and offer ways to partner on some of the solutions, we can move the agenda a long way. The biggest part is just to start playing our role.

MJ: What do you think of this notion of an environmental czar, either in America or abroad?

TL: We have one. Carol Browner is the environment and energy czar in the White House. But it doesn't work unless you have a president who understands it. The really great news is that Carol said to me, "Not all of the administration gets it, but the president really gets it." That's key.

MJ: From my perspective as an educator, the big thing now is to teach sustainability at all different levels. Last year we gathered a large group of top thinkers to consider what it means to create a curriculum based on sustainable education. Some folks said we should spend two more years in a school of forestry or studying ecology on top of whatever other degree you are doing. Others said that reading some biology books is not going to make you an expert, or really help whatever other field you are working in.

TL: The solutions are not going to come just from science or just from economics or just from design. They'll derive from a certain common understanding of that whole suite of things. It's important to have enough basic science so that you don't make mistakes with all the best intentions, but I think poets can make as much of a contribution as biologists. Even lawyers can contribute.

MJ: Ha!

TL: But seriously, some of our important environmental leaders have been lawyers.

MJ: I'm finding that a lot of people in education resist the sustainability movement because they find it a bit preachy. It's loaded with virtues that haven't been questioned enough. We just seem to accept them as bromides or axiomatic statements.

TL: It is so easy for anybody concerned about these things to become self-righteous without even intending to. So while part of me would like to say "that's right" and "that's wrong," we have to be really careful not to caricature ourselves. So my approach is to say, "I'm in the problem-solving game."

Which is how I came up with the idea of debt-for-nature swaps. It was the late seventies, when many Latin American nations had enormous debts to commercial banks. I happened to be at a hearing about the environmental impacts of the World Bank when suddenly one of the witnesses started going off about the social impacts driven by debt. I'm sitting there listening, and I realized that debt itself was causing environmental problems.

So I started thinking, maybe there's a way to turn that around, to actually use that debt for some solutions. That's basically what I proposed, and there have been many different forms of it ever since. The United States, for example, will forgive the debt of, say, Ecuador for X million dollars if Ecuador agrees to spend an equivalent amount of local currency on a mutually agreed upon environmental program.

MJ: But critics say these resources are being put aside in developing countries where local people can't access them.

TL: The justified critique is not about the mechanism; it's about the individual project. The very first one in Bolivia, for instance, was done so quickly that the local people really weren't engaged, and it created a backlash. It's simply a financial mechanism—if we have so many problems on the table, let's think about them as things to be solved rather than preaching.

MJ: Right. John F. Kennedy said, "Man created problems; man can solve problems."

TL: Exactly.

MJ: So I want to bring up something controversial. I don't agree with this man, I don't even know if I should say his name, but he's written a book called *The Skeptical Environmentalist*. At some level, I will applaud one thing that he's done, that he made people more critical of statistics—

TL: It's actually very, very bad scholarship. The publisher's review process had to have broken down. I happen to have just reviewed a new book where the author begins with something I said in the first review of *The Skeptical Environmentalist*. I'd said that time and time again, I would go to look for the original citation to discover nothing but a mirage in the desert.

That said, your point is that it's always incumbent upon us to be very careful about our facts and our figures.

MJ: Yes. But also what I'm getting at is that there needs to be a little more conflict—to make things not contentious, but self-critiquing. I'd expect that there are eight-hundred-something papers that prove that climate change is happening and that seem to be in almost universal agreement that we accept this is an absolute truth. Maybe we have to figure out a baseline or fundamental way of critiquing them.

I'll make a more concrete case: diapers. DuPont did a life cycle analysis on baby diapers some time ago. They had some really good scientists who produced a peer-review document that said to the public, plastic diapers are better than cloth diapers. Pound for pound, the amount of carbon, water, and electricity consumed is less if we use plastic.

Thomas E. Lovejoy and Mitchell Joachim

But a couple of years later, someone went through the numbers and found a mistake. Not an intentional mistake, but a mistake. If you looked at the equation about drying cloth diapers, it assumed you put them in a dryer, using a lot of electricity, right?

TL: Put them on the clothesline.

MJ: Exactly. You hang them out to dry. It was a totally shit scam—this message that was sent out from the realm of science misled the public and dumped tons of plastic into the environment.

TL: My reaction to that is really twofold. One is that's actually how science works.

MJ: That's true.

TL: It tends to be self-correcting, because it is self-criticizing. My other question is, What are you going to do with your baby?

MJ: Design a new kind of diaper—"soft diapers?"

TL: I couldn't resist. Your wife is weeks away, right?

MJ: Yes, weeks away. ∞

Acknowledgments

My deep gratitude is to the many people who've had a hand in inspiring, producing, or contributing to the Seed Salon. Laura McNeil, Sari Globerman, and Maywa Montenegro, who edited the Seed Salon and brought to a nascent department their passion and curiosity; our editors, who took a vision for a magazine and made it one; Julian Dufort, who elegantly captured the people behind the ideas in this book with his photography; Peter Hubbard, our book editor, and Max Brockman, our literary agent, who made this project a reality; John Brockman and Edward O. Wilson, whose writings deeply inspired me to start *Seed*; Laszlo Kato, my neighbor growing up in Montreal, who taught me what science is; Paola Antonelli and Stefan Sagmeister, who taught me what design is; Yofi Sadaka, Pauline Gagnon, Carole Charlebois, and Maureen O'Connor-McCourt, who gave me the opportunities to fall in love with science; Jessica Banks, Ari Wallach, and Eva Wisten, whose support has been invaluable; my family, whose love and honesty has kept me sane; Ayah Bdeir, who means everything to me. Finally, all our friends who participated in the Seed Salon—Paola Antonelli, Albert-László Barabasi, David Byrne, Noam Chomsky, Richard Colton, Laurie David, Daniel C. Dennett, Drew Endy, Joan Fontcuberta, James Fowler, Peter Galison, Michael Gazzaniga, Rebecca Goldstein, Michel Gondry, Marc Hauser, Lynn Hershman Leeson, Chuck Hoberman, Natalie Jeremijenko, Mitchell Joachim, Lawrence Krauss, Jonathan Lethem, Janna Levin, Daniel Levitin, Alan Lightman, Thomas Lovejoy, Benoit Mandelbrot, Errol Morris, Steven Pinker, Lisa Randall, Carlo Ratti, Ariel Ruiz i Altaba, Stefan Sagmeister, Stephen Schneider (who sadly passed away in July 2010), Will Self, Michael Shanks, Paul Steinhardt, Robert Stickgold, Steven Strogatz, Jill Tarter, Robert Trivers, Spencer Wells, Edward O. Wilson, Tom Wolfe, and Will Wright—and without whom *Seed* would be nothing.

Index